Spring Boot 3.0 Cookbook

Proven recipes for building modern and robust Java
web applications with Spring Boot

Felip Miguel Puig

Spring Boot 3.0 Cookbook

Group Product Manager: Kaustubh Manglurkar

Publishing Product Manager: Bhavya Rao

Book Project Manager: Shagun Saini

Senior Editor: Debolina Acharyya

Technical Editor: Reenish Kulshrestha

Copy Editor: Safis Editing

Indexer: Subalakshmi Govindhan

Production Designer: Jyoti Kadam

DevRel Marketing Coordinators: Anamika Singh and Nivedita Pandey

First published: July 2024

Production reference: 1130624

Published by Packt Publishing Ltd.

Grosvenor House

11 St Paul's Square

Birmingham

B3 1RB, UK

ISBN 978-1-83508-949-1

www.packtpub.com

To my wife, Àgata Font; you and I make the best team imaginable. To Rita, Telma, and Liam; thank you for teaching me life's true priorities. To my mother, Pepita, and the memory of my father, Gregorio; I am grateful for the values you instilled in me.

– Felip Miguel Puig

Contributors

About the author

Felip Miguel Puig has been working in the technology industry since 2000, particularly interested in software development and Java and C#. He is currently a principal software engineer at Dynatrace, a leading company in the monitoring and observability industry. Before joining Dynatrace, Felip worked at Microsoft for 17 years in various engineering and consulting roles, where he delivered noteworthy projects for significant clients in different industries. He is an expert in cloud technology and has been working in this field since 2012.

I want to thank my wife, Àgata, for her support and patience with my geek hobbies, such as writing this book.

About the reviewers

Romit Sutariya is a software engineer skilled in enhancing applications by identifying areas for improvement, providing recommendations, and implementing solutions. He has successfully managed multiple projects utilizing Java, Spring Boot, microservices, AWS, and Kubernetes and adhering to clean coding practices for optimal quality and performance. His educational background includes an MSc in computer and information technology. Additionally, Romit holds certifications for AWS Solutions Architect Associate and Kubernetes, showcasing his dedication to continuous learning and applying new technologies to solve real-world challenges.

Ibidapo Abdulazeez is currently a master's student at York St John University with a focus on software engineering. Prior to embarking on his graduate journey, he accumulated two years of invaluable work experience in the field of software engineering. During this time, he had the opportunity to architect, design, and develop enterprise solutions. His other areas of expertise are **Amazon Web Services (AWS)**, Azure, and databases, to mention a few.

Table of Contents

2

Securing Spring Boot Applications with OAuth2   35

3

Observability, Monitoring, and Application Management   91

Part 2: Database Technologies

5

Data Persistence and Relational Database Integration with Spring Data   187

6

Data Persistence and NoSQL Database Integration with Spring Data   243

Part 3: Application Optimization

7

Finding Bottlenecks and Optimizing Your Application   287

8

Spring Reactive and Spring Cloud Stream 317

Part 4: Upgrading to Spring Boot 3 from Previous Versions

9

Upgrading from Spring Boot 2.x to Spring Boot 3.0 357

Preface

Spring Boot is Java's most popular framework for web and microservices development. It allows you to create production-grade applications with minimal configuration, following its "convention over configuration approach."

Spring Boot is always evolving and adapting to the latest technology trends. Its ecosystem allows you to integrate with any technology, from databases to AI, and use traversal features such as observability and security. You can use it for virtually any type of application.

In this book, we'll cover the most common scenarios in a hands-on way, and you'll learn the foundations to use the vast features available.

Who this book is for

This book is for Java developers who want to gain expertise in modern web development, architects designing complex systems, experienced Spring Boot developers and technology enthusiasts looking to stay up to date with the latest trends, and software engineers in need of practical solutions for everyday challenges. Hands-on experience with Java is required. Prior development experience on the cloud will be useful, but not necessary.

What this book covers

Chapter 1, *Building RESTful APIs*, teaches you how to write, consume, and test RESTful API with Spring Boot 3.

Chapter 2, *Securing Spring Boot Applications with OAuth2*, shows you how to deploy an authorization server and use it to protect a RESTful API and a website. You will learn how to authenticate users with Google accounts. You will also learn how to protect applications with Azure AD B2C.

Chapter 3, *Observability, Monitoring, and Application Management*, explores how to leverage the observability features available in Spring Boot with Actuator. This chapter uses Open Zipkin, Prometheus, and Grafana, consuming the observability data exposed by a Spring Boot application for monitoring.

Chapter 4, *Spring Cloud,* covers how to use Spring Cloud to develop a distributed system composed of several microservices. We'll use Eureka Server, Spring Cloud Gateway, Spring Config, and Spring Boot Admin.

Chapter 5, Data Persistence and Relational Database Integration with Spring Data, delves into how to integrate an application with PostgreSQL using Spring Data JPA. You will define repositories and use **Java Persistance Query Language** (**JPQL**) and Native SQL. You will learn to use transactions, database versioning, and migrations with Flyway, and Testcontainers for integration tests.

Chapter 6, Data Persistence and NoSQL Database Integration with Spring Data, explains the benefits and trade-offs of using NoSQL databases such as MongoDB and Cassandra, teaching you how to tackle some common challenges of NoSQL databases, such as data partitioning or concurrency management. You will use Testcontainers for integration tests with MongoDB and Cassandra.

Chapter 7, Finding Bottlenecks and Optimizing Your Application, describes how to run a load test using JMeter against a Spring Boot application, apply different optimizations, such as caching or building a native application, and compare the improvements with the original results. You will also learn some useful techniques to prepare a Native application, such as using the GraalVM Tracing Agent.

Chapter 8, Spring Reactive and Spring Cloud Stream, explores how to use Spring Reactive for high concurrency scenarios, creating a reactive RESTful API, a reactive API client, and the R2DBC driver for PostgreSQL. You will learn how to create an event-driven application, using Spring Cloud Stream connected to a RabbitMQ server.

Chapter 9, Upgrading from Spring Boot 2.x to Spring Boot 3.0, explains how to manually upgrade a Spring Boot 2.6 application to the latest version of Spring Boot 3. You will prepare the application before upgrading to Spring Boot 3 and fix all the issues, step by step, after the upgrade. You will also learn how to use OpenRewrite to automate part of the migration process.

To get the most out of this book

You will need the JDK 21 for all the chapters of this book. In *Chapter 9*, you will also need JDK 11 and JDK 17. I recommend using a tool such as SDKMAN! to install and configure the SDK on your computer. If you use Windows, you can use the JDK installer.

I used Maven as a dependency and build system for all samples. You can optionally install it on your computer, but all projects created in this book use the Maven Wrapper, which downloads all dependencies if needed.

If you are a Windows user, I recommend using **Windows Subsystem for Linux** (**WSL**), as some of the complementary tools used in this book are available in Linux, and the scripts available in the book's GitHub repository are tested in Linux only. Indeed, I'm a Windows user and used WSL for all the samples prepared for this book.

I also recommend installing Docker, as it's the simplest way to run some of the services integrated with this book, such as PostgreSQL. Docker is the best option to run a distributed system, composed of different applications talking to each other on your computer. In addition, most of the integration tests use Testcontainers, which requires Docker.

I tried to explain all samples without specific IDE requirements in this book. I used Visual Studio Code, primarily for its excellent integration with WSL, but you can use any other IDE of your preference, such as IntelliJ or Eclipse.

Software/hardware covered in the book	OS requirements
OpenJDK 21	Windows, macOS, or Linux
OpenJDK 11 and 17	Windows, macOS, or Linux
Docker	Windows (recommended with WSL integration), macOS, or Linux
Prometheus	On Docker (recommended) or natively on Windows, macOS, or Linux
Grafana	On Docker (recommended) or natively on Windows, macOS, or Linux
OpenZipkin	On Docker (recommended) or natively running Java on Windows, macOS, or Linux
PostgreSQL	On Docker (recommended) or natively on Windows, macOS, or Linux.
MongoDB	On Docker (recommended) or natively on Windows, macOS, or Linux
Apache Cassandra	On Docker (recommended) or natively on Linux
RabbitMQ	On Docker (recommended) or natively on Windows, macOS, or Linux
JMeter	Windows, macOS, or Linux
An IDE such as Visual Studio Code/IntelliJ	Windows, macOS, or Linux

If you are using the digital version of this book, we advise you to type the code yourself or access the code via the GitHub repository (link available in the next section). Doing so will help you avoid any potential errors related to the copying and pasting of code.

Download the example code files

You can download the example code files for this book from GitHub at `https://github.com/PacktPublishing/Spring-Boot-3.0-Cookbook`. If there's an update to the code, it will be updated on the existing GitHub repository. Some recipes use the previous recipes as the starting point. In those cases, I provide a working version in the `start` subfolder of each recipe and the complete version in the `end` folder.

We also have other code bundles from our rich catalog of books and videos available at `https://github.com/PacktPublishing/`. Check them out!

Conventions used

There are a number of text conventions used throughout this book.

`Code in text`: Indicates code words in text, database table names, folder names, filenames, file extensions, pathnames, dummy URLs, user input, and Twitter handles. Here is an example: "This behavior can be configured using the propagation attribute of the `@Transactional` annotation."

A block of code is set as follows:

```
em.getTransaction().begin();
// do your changes
em.getTransaction().commit();
```

When we wish to draw your attention to a particular part of a code block, the relevant lines or items are set in bold:

```
public int getTeamCount() {
    return jdbcTemplate.queryForObject("SELECT COUNT(*) FROM teams",
Integer.class);
}
```

Any command-line input or output is written as follows:

```
docker-compose -f docker-compose-redis.yml up
```

Bold: Indicates a new term, an important word, or words that you see on screen. For example, words in menus or dialog boxes appear in the text like this. Here is an example: "If you run the application now, you'll see a new queue named **match-events-topic.score.dlq**."

> **Tips or important notes**
> Appear like this.

Sections

In this book, you will find several headings that appear frequently (*Getting ready, How to do it..., How it works..., There's more...,* and *See also*).

To give clear instructions on how to complete a recipe, use these sections as follows.

Getting ready

This section tells you what to expect in the recipe and describes how to set up any software or any preliminary settings required for the recipe.

How to do it...

This section contains the steps required to follow the recipe.

How it works...

This section usually consists of a detailed explanation of what happened in the previous section.

There's more...

This section consists of additional information about the recipe in order to make you more knowledgeable about the recipe.

See also

This section provides helpful links to other useful information for the recipe.

Get in touch

Feedback from our readers is always welcome.

General feedback: If you have questions about any aspect of this book, mention the book title in the subject of your message and email us at customercare@packtpub.com.

Errata: Although we have taken every care to ensure the accuracy of our content, mistakes do happen. If you have found a mistake in this book, we would be grateful if you would report this to us. Please visit www.packtpub.com/support/errata, select your book, click on the **Errata Submission Form** link, and enter the details.

Piracy: If you come across any illegal copies of our works in any form on the Internet, we would be grateful if you would provide us with the location address or website name. Please contact us at copyright@packtpub.com with a link to the material.

If you are interested in becoming an author: If there is a topic that you have expertise in and you are interested in either writing or contributing to a book, please visit authors.packtpub.com.

Share Your Thoughts

Once you've read *Spring Boot 3.0 Cookbook*, we'd love to hear your thoughts! Scan the QR code below to go straight to the Amazon review page for this book and share your feedback.

https://packt.link/r/1835089496

Your review is important to us and the tech community and will help us make sure we're delivering excellent quality content.

Download a free PDF copy of this book

Thanks for purchasing this book!

Do you like to read on the go but are unable to carry your print books everywhere?

Is your eBook purchase not compatible with the device of your choice?

Don't worry, now with every Packt book you get a DRM-free PDF version of that book at no cost.

Read anywhere, any place, on any device. Search, copy, and paste code from your favorite technical books directly into your application.

The perks don't stop there, you can get exclusive access to discounts, newsletters, and great free content in your inbox daily

Follow these simple steps to get the benefits:

1. Scan the QR code or visit the link below

https://packt.link/free-ebook/9781835089491

2. Submit your proof of purchase
3. That's it! We'll send your free PDF and other benefits to your email directly

Part 1:
Web Applications
and Microservices

In this part, we cover the basics of RESTful APIs, distributed applications, and microservices with Spring Cloud, as well as cross-functional features such as security and observability.

This part has the following chapters:

- *Chapter 1, Building RESTful APIs*
- *Chapter 2, Securing Spring Boot Applications with Oauth2*
- *Chapter 3, Observability, Monitoring, and Application Management*
- *Chapter 4, Spring Cloud*

1

Building RESTful APIs

RESTful APIs are crucial in modern cloud apps for seamless data exchange, enabling interoperability, scalability, and efficient communication between services. Spring Boot simplifies RESTful authoring by providing a framework for quick, efficient development, auto-configuration, and integrated tools.

In this chapter, you will acquire the skills to create RESTful services and consume them seamlessly from other applications. You will also learn how to create automated tests for your RESTful APIs using the features provided by Spring Boot and other popular tools.

In this chapter, we're going to cover the following main recipes:

- Creating a RESTful API
- Defining responses and the data model exposed by the API
- Managing errors in a RESTful API
- Testing a RESTful API
- Using OpenAPI to document our RESTful API
- Consuming a RESTful API from another Spring Boot application using FeignClient
- Consuming a RESTful API from another Spring Boot application using RestClient
- Mocking a RESTful API

Technical requirements

To complete this chapter's recipes, you will need a computer with any OS (I use Ubuntu on Windows Subsystem for Linux – WSL), an editor such as Visual Studio Code (https://code.visualstudio.com/) or IntelliJ Idea (https://www.jetbrains.com/idea/), and Java OpenJDK 17 or higher.

There are multiple distributions of Java from different vendors – if you already have one installed, you can continue using it; if you need to install one, you can use Eclipse Adoptium distribution (https://adoptium.net/).

If you use Visual Studio Code, I recommend installing Extension Pack for Java (`https://marketplace.visualstudio.com/items?itemName=vscjava.vscode-java-pack`) and Spring Boot Extension Pack (`https://marketplace.visualstudio.com/items?itemName=vmware.vscode-boot-dev-pack`).

If you don't have a tool to perform HTTP requests, you could use curl (`https://curl.se/`) or Postman (`https://www.postman.com/`).

Finally, you can download the complete project on GitHub at: `https://github.com/PacktPublishing/Spring-Boot-3.0-Cookbook/`.

You will need a git client to download the code from the book's GitHub repository (`https://git-scm.com/downloads`).

Creating a RESTful API

A **RESTful API** is a standardized way for software components to communicate over the internet using HTTP methods and URLs. You should learn it because it's fundamental for modern web and cloud application development. It promotes scalable, flexible, and stateless communication, enabling developers to design efficient and widely compatible systems. Understanding RESTful APIs is crucial for building and integrating services.

When I was a child I played with football-player cards, exchanging cards that I had multiples of with my friends. My children, some decades later, still play this game. Throughout this chapter, you will create a system to manage a football card-trading game, with teams, players, albums, and cards. In this recipe, you will create a RESTful API exposing **Create, Read, Update, and Delete** (**CRUD**) **operations** on football players.

Getting ready

To create a RESTful API, Spring Boot provides a great tool named Spring Initializr. You can open this tool in your browser using `https://start.spring.io/`. We'll use this tool to create a Spring Boot project with all dependencies. This tool is also well integrated into code editors such as VSCode and IntelliJ (Premium edition).

How to do it...

Let's create a RESTful project using Spring Initializr and create our first endpoint with typical HTTP operations:

1. Open https://start.spring.io in your browser and you'll see the following screen:

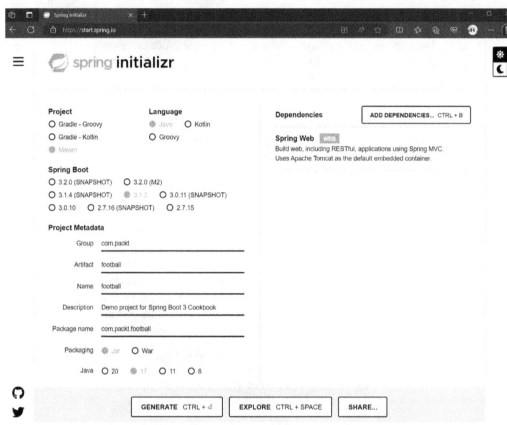

Figure 1.1: Spring Initializr

Spring Initilizr allows you to configure the project and generates the structure with the necessary files and dependencies, which you can use as a starting point for your application. On this start page, set up the following configuration:

- In the **Project** section, select **Maven**

- In the **Language** section, select **Java**

- In the **Spring Boot** section, select the latest stable version – at the time of writing this book, this is **3.1.4**

- In the **Dependencies** section, select **Spring Web**
- Do the following in the **Project Metadata** section:
  - For **Group**, type com.packt
  - For **Artifact**, type football
  - The **Name** and **Package name** will be autogenerated from previous fields, so keep them as is
  - In the **Description** field, type a description of this project, such as Demo project for Spring Boot 3 Coobook
  - For **Packaging**, select **Jar**
  - For **Java**, select **21**

2. Once you have configured the preceding options, you can choose the options for **Generate**, **Explore**, or **Share…**:

   - If you click **Explore**, you can explore the project before downloading it.
   - If you click **Share**, a URL is generated that you can share with other people. For instance, our configuration will produce this URL: https://start.spring.io/#!type=maven- project &language=java&platformVersion=3.1.3&packaging=jar&jvmVersion=1 7&groupId=compackt&artifactId=football&name=football& description=Demo%20project%20for%20Spring%20Boot%203%20 Cookbook&packageName=c om.packt.football&dependencies=web. If you open it, it will configure all of the options as shown in *Figure 1.1*.
   - If you click **Generate**, it will download a ZIP file of the project structure. Click this option now.

3. Unzip the file. You now have the basic project structure, but you don't have any API. If you try to run the application, you will receive an HTTP 404 Not Found response.

4. In the src/main/java/com/packt/football folder, create a file named PlayerController.java with the following content to create a RESTful endpoint:

```
package com.packt.football;
import java.util.List;
import org.springframework.web.bind.annotation.*;
@RequestMapping("/players") @RestController
public class PlayerController {
    @GetMapping
    public List<String> listPlayers() {
        return List.of("Ivana ANDRES", "Alexia PUTELLAS");
    }
}
```

5.   To run it, open a terminal in the project root folder and execute the following command:

```
./mvnw spring-boot:run
```

This command will build your project and start the application. By default, the web container listens on port 8080.

6.   Execute an HTTP request to see the results. You can open `http://localhost:8080/players` in a browser or use a tool such as curl to perform the request:

```
curl http://localhost:8080/players
```

You will have a list of players returned by the controller.

7.   Enhance your RESTful endpoint by adding more verbs. In the `PlayerController.java` file, do the following:

- Implement a POST request to create a player:

```
@PostMapping
public String createPlayer(@RequestBody String name) {
    return "Player " + name + " created";
}
```

- Add another GET request to return one player:

```
@GetMapping("/{name}")
public String readPlayer(@PathVariable String name) {
    return name;
}
```

- Add a DELETE request to delete one player:

```
@DeleteMapping("/{name}")
public String deletePlayer(@PathVariable String name) {
    return "Player " + name + " deleted";
}
```

- Implement a PUT request to update a player:

```
@PutMapping("/{name}")
public String updatePlayer(@PathVariable String name, @RequestBody String newName) {
    return "Player " + name + " updated to " + newName;
}
```

8. Execute your application again as explained in *step 5* and test your endpoint. Perform a GET request by entering the following command into your terminal:

```
curl http://localhost:8080/players/Ivana%20ANDRES
```

And you will receive this output:

```
Ivana ANDRES
```

Perform a POST request using curl:

```
curl --header "Content-Type: application/text" --request POST
--data 'Itana BONMATI' http://localhost:8080/players
```

And you will receive this output:

Player Itana BONMATI created

Perform a PUT request using curl:

```
curl --header "Content-Type: application/text" --request PUT
--data 'Aitana BONMATI' http://localhost:8080/players/Itana%20
BONMATI
```

And you will receive this output:

```
Player Itana BONMATI updated to Aitana BONMATI
```

Perform a DELETE request:

```
curl --header "Content-Type: application/text" --request DELETE
http://localhost:8080/players/Aitana%20BONMATI
```

And you will receive this output:

```
Player Aitana BONMATI deleted
```

How it works...

By adding the Spring Web dependency to our project, Spring Boot automatically embeds a Tomcat server in the application. **Tomcat** is an open source web server and servlet container developed by the Apache Software Foundation. It's one of the most popular choices for hosting Java-based web applications and services. You can run the application right after downloading it from start. spring.io and it listens on port 8080, which is the default port for Tomcat. However, as there is no mapping configured in the application, it always responds with a 404 Not Found error.

By adding the **@RestController** annotation to the PlayerController class, we are informing Spring Boot that it should register the PlayerController class in dependency containers as an implementation class. By adding **@RequestMapping**, we inform the web container to map HTTP requests to their handler, in this case, the PlayerController class. As we applied these at a class level, we configured all requests in this class with the players prefix.

The last step is mapping requests to their handling methods. This is performed by using mapping annotations:

- `@GetMapping`: Maps a GET request to a method
- `@PostMapping`: Maps a POST request to a method
- `@PutMapping`: Maps a PUT request to a method
- `@DeleteMethod`: Maps a DELETE request to a method

These mapping annotations are a specialization of `@RequestMapping`, as they inform the web container how to map a request to its handler, in this case, using the annotated method.

Keep in mind that you can only have one of these mapping types per controller (i.e., one `GetMapping` or `PostMapping`) unless you provide more configuration to refine the mapping. In this example, you can see there are two instances of `@GetMapping`, but `readPlayer` is annotated with an additional element and thus is mapped with its class prefix, `players`, plus a name. That means all GET requests with `/players/anything` will be mapped to this method.

As of now, that additional information is not yet configured in the method. To use all this additional HTTP request information in your method, you can use the following annotations:

- **@PathVariable**: This will map one part of the HTTP request path to the method argument. For instance, `@GetMapping("/{name}") public String readPlayer(@ PathVariable String name)` maps the last part of the path to the name method argument.
- **@RequestBody**: This will map the request body to the method argument.
- **@RequestHeader**: This will map request headers to the method argument.
- **@RequestParam**: You can use this to map requests params, such as query string parameters, form data, or parts in multipart requests.

Just by decorating our classes with previous annotations, Spring Boot is able to set up the web application container to manage the requests. There are some annotations not yet covered, but we covered the basics to create our RESTful APIs.

There's more...

Even if the RESTful endpoint we just created is very simple, I intentionally added these methods – GET by default, GET with an identifier, POST, PUT, and DELETE. This choice is rooted in the fact that it aligns with the most prevalent semantics for performing CRUD and List operations on a resource.

In the context of our resource being football *players*, we have the following operations:

- GET by default usually returns a list of resources, in our case, all players
- GET with an identifier returns a specific player

- POST creates a new resource
- PUT updates a resource
- DELETE deletes a resource

Also, the HTTP status code responses are very important in the semantics of RESTful operations. In this recipe, the responses are not managed in the standard way. In the following recipes, we will expand on this and learn how Spring Boot can facilitate the proper handling of responses.

See also

If you want to learn more about API design, you can visit the following pages:

- `https://swagger.io/resources/articles/best-practices-in-api-design/`
- `https://learn.microsoft.com/azure/architecture/best-practices/api-design`

Defining responses and the data model exposed by the API

In the previous recipe, we created a very simple RESTful API. To develop a RESTful API that provides a positive user experience for its consumers, it is essential to incorporate both standard response codes and a consistent data model. In this recipe, we will enhance the previous RESTful API by returning standard response codes and creating a data model for our players endpoint.

Getting ready

You can use the project generated in the previous recipe or download the sample from the GitHub repository: `https://github.com/PacktPublishing/Spring-Boot-3.0-Cookbook/`.

You can find the code to start this exercise in the `chapter1/recipe1-2/start` folder.

How to do it...

In this recipe, we will create a folder structure to contain different types of classes for our project. We will define a data model to expose in our RESTful API, along with a service to provide the operations needed by the API.

Note that all of the content created in the following steps will be under the `src/main/java/com/packt/football` folder or one of the subfolders you will create when required. Let's get started:

1. Create a folder named `model`. Then in this folder, create a file named `Player.java` with the following content:

    ```java
    public record Player(String id, int jerseyNumber, String name,
    String position,     LocalDate dateOfBirth) {
    }
    ```

2. Create a folder named `exceptions`. In this folder, create two files:

    *   The first, `AlreadyExistsException.java`, should have the following content:

    ```java
    package com.packt.football.exceptions;
    public class AlreadyExistsException extends RuntimeException {
        public AlreadyExistsException(String message) {
            super(message);
        }
    }
    ```

    *   And the second, `NotFoundException.java`, should contain this content:

    ```java
    package com.packt.football.exceptions;
    public class NotFoundException extends RuntimeException {
        public NotFoundException(String message) {
            super(message);
        }
    }
    ```

3. Create another folder named `services`, and in this folder, create a class named `FootballService`. This class manages all operations needed by our RESTful API. Do the following in this file:

    *   Create the class first:

    ```java
    @Service
    public class FootballService {
    }
    ```

    This class will manage the data in a Map, keeping all players in memory.

    *   Now let's define a `Map<String, Player>` field and initialize it. (I just created two entries for brevity, but in the GitHub repo, you will find many more):

    ```java
    private final Map<String, Player> players = Map.ofEntries(
            Map.entry("1884823", new Player("1884823", 5, "Ivana
    ANDRES", "Defender", LocalDate.of(1994, 07, 13))),
    ```

```
        Map.entry("325636", new Player("325636", 11, "Alexia
PUTELLAS", "Midfielder", LocalDate.of(1994, 02, 04
)))));
```

- Define the operations required by our RESTful API:

  - Start by listing the players:

```
public List<Player> listPlayers() {
    return players.values().stream()
            .collect(Collectors.toList());
}
```

  - Then return a player (do note that if the player doesn't exist, it will throw an exception):

```
public Player getPlayer(String id) {
    Player player = players.get(id);
     if (player == null)
        throw new NotFoundException("Player not found");
   return player;
}
```

  - Add a new player (do note that if the player already exists, it will throw an exception):

```
public Player addPlayer(Player player) {
    if (players.containsKey(player.id())) {
        throw new AlreadyExistsException("The player already
exists");
    } else {
        players.put(player.id(), player);
        return player;
    }
}
```

  - Update a player (note that if the player doesn't already exist, it will throw an exception):

```
public Player updatePlayer(Player player) {
    if (!players.containsKey(player.id())) {
        throw new NotFoundException("The player does not
exist");
    } else {
        players.put(player.id(), player);
       return player;
    }
}
```

- Delete a player (note that if the player doesn't exist, it will continue without errors):

```
public void deletePlayer(String id) {
    if (players.containsKey(id)) {
        players.remove(id);
    }
}
```

4. Next, in the `PlayerController` class, modify the controller to use our new service and expose the newly created data model:

- Add a `FootballService` field and create a constructor in the `PlayerController` class with an argument of type `FootballService` to initialize it:

```
@RequestMapping("/players")
@RestController
public class PlayerController {
    private FootballService footballService;


    public PlayerController(FootballService footballService) {
        this.footballService = footballService;
    }
}
```

- Create the operations to manage the players. We will use our recently created service to manage that functionality. As explained in the previous recipe, we will decorate our class methods to manage the RESTful endpoint methods and will invoke the services of our `Football` service class:

```
@GetMapping
public List<Player> listPlayers() {
    return footballService.listPlayers();
}
@GetMapping("/{id}")
public Player readPlayer(@PathVariable String id) {
    return footballService.getPlayer(id);
}
@PostMapping
public void createPlayer(@RequestBody Player player) {
    footballService.addPlayer(player);
}
@PutMapping("/{id}")
public void updatePlayer(@PathVariable String id,
                          @RequestBody Player player) {
```

```
        footballService.updatePlayer(player);
    }
    @DeleteMapping("/{id}")
    public void deletePlayer(@PathVariable String id) {
        footballService.deletePlayer(id);
    }
```

5. In the `application` root folder, open a terminal and execute the following command to run the application:

    ```
    ./mvnw spring-boot:run
    ```

6. Test the application by executing the following `curl` command to get all of the players:

    ```
    curl http://localhost:8080/players
    [{"id":"325636","jerseyNumber":11,"name":"Alexia
    PUTELLAS","position":"Midfielder","dateOfBirth":"1994-02-
    04"},{"id":"1884823","jerseyNumber":5,"name":"Ivana
    ANDRES","position":"Defender","dateOfBirth":"1994-07-13"}]
    ```

How it works...

In this recipe, we defined a new record type named `Player`. Spring Boot automatically serializes that object into a response body that can be sent to the client in a format such as JSON or XML.

Spring Boot uses a message converter to perform this serialization. The choice of message converter and the serialization format depends on the Accept header in the client's request. By default, Spring Boot serializes the response as JSON.

> **About records**
>
> The **record** feature was introduced in Java 16. Java records provide a convenient way to declare classes that are simple data carriers, automatically generating methods such as the `equals()`, `hashCode()`, and `toString()` constructors based on the record components. This feature aims to simplify the creation of classes that primarily encapsulate data. Spring Boot 3 uses Java 17 or higher.

If you have special serialization requirements you can configure your own message converter by implementing your `WebMvcConfigurer` and overriding the method `configureMessageConverters`. You can find more information in the Spring Framework documentation: `https://docs.spring.io/spring-framework/docs/current/javadoc-api/org/springframework/web/servlet/config/annotation/WebMvcConfigurer.html#configureMe%20ssageConverters(java.util.List)`.

Spring Boot's default handling of HTTP status codes can be summarized as follows:

- When the execution occurs without generating exceptions, it responds with an HTTP 200 status.

- If a method is not implemented by the endpoint, it will return a **405 Method not allowed** error.

- If trying to get a resource that doesn't exist, for instance, a path that is not managed by the application, it will return **404 Not found**.

- If the request is not valid, it will return **400 Bad Request**.

- In the event of an exception, it yields an **HTTP 500 Internal Server error**.

- There are other operations related to security, which we will discuss in later chapters, that may return **401 Unauthorized** or **403 Forbidden**.

This behavior can be enough in some scenarios, but if you want to provide proper semantics to your RESTful API, you should return a 404 status code when a resource is not found. Check the next recipe to learn how to handle these scenarios.

Note that the `FootballService` class is annotated with **@Service**. That registers the class as a Spring bean and hence it is available in the Inversion-of-Control container. When a Spring Boot application starts, it scans for classes annotated with various stereotypes, such as `@Service`, `@Controller`, `@Bean`, and others. As there is a dependency in the `PlayerController` class for `FootballService`, when Spring Boot instantiates the `PlayerController`, it passes an instance of `FootballService` class.

Managing errors in a RESTful API

In the previous recipe, we enhanced our RESTful API by using complex data structures. However, the application was not able to manage some common errors or return standard response codes. In this recipe, l we will enhance the previous RESTful API by managing common errors and returning consistent response codes following the standards.

Getting ready

You can use the project generated in the previous recipe or download the sample from the GitHub repository at `https://github.com/PacktPublishing/Spring-Boot-3.0-Cookbook/`.

You can find the code to start this exercise in the `chapter1/recipe1-3/start` folder.

How to do it...

In this recipe, we will modify the RESTful API created in the previous recipe to handle the exceptions that can be raised by our application and we'll return the most appropriate HTTP response code.

All content created in the following steps will be under the `src/main/java/com/packt/football` folder or one of the subfolders you will create. Let's get started:

1. If you try to retrieve a non-existing player or create the same player twice, it will throw an exception. The result will be an HTTP 500 server error:

```
curl http://localhost:8080/players/99999
{"timestamp":"2023-09-
16T23:18:41.906+00:00","status":500,"error":"Internal Server
Error","path":"/players/99999"}
```

2. To manage this error more consistently, we will add a new `notFoundHandler` method in the `PlayerController` class to manage `NotFoundException` errors:

```
@ResponseStatus(value = HttpStatus.NOT_FOUND, reason = "Not
found")
@ExceptionHandler(NotFoundException.class)
public void notFoundHandler() {
}
```

3. Next, we'll add another method named `alreadyExistsHandler` to manage `AlreadyExistsException` errors:

```
@ResponseStatus(value = HttpStatus.BAD_REQUEST, reason =
"Already exists")
@ExceptionHandler(AlreadyExistsException.class) public void
alreadyExistsHandler() {
}
```

4. In the `application` root folder, open a terminal and execute the following command to run the application:

```
./mvnw spring-boot:run
```

5. Test the application by executing the following curl commands:

   - Execute this command to get a player that does not exist:

```
curl http://localhost:8080/players/99999
{"timestamp":"2023-09-
16T23:21:39.936+00:00","status":404,"error":"Not
Found","path":"/players/99999"}
```

   - Note that by returning an HTTP 404 Not Found response, our application adheres to standard RESTful API semantics. HTTP 404 means that you tried to get a resource that does not exist, in our case, player 9999.

- Let's verify that our application manages the `AlreadyExistsException` as expected. Execute the following request to create a player twice:

```
data="{'id': '8888', 'jerseyNumber':6, 'name':'Cata COLL',"
data=${data}" 'position':'Goalkeeper', "
data=${data}" 'dateOfBirth': '2001-04-23'}"
curl --header "Content-Type: application/json" --request POST \
  --data $data  http://localhost:8080/players
```

The first time it will work with no errors and will return the HTTP 200 code. The second time it will return an HTTP 400 code.

How it works...

As we learned in the previous recipe, Spring Boot manages the HTTP status code for the most common cases. In this recipe, we demonstrated how to manage other scenarios that are specific to our application logic and require consistent HTTP status codes.

To provide proper semantics to the RESTful API, you should return a 404 status code when a resource is not found. In some scenarios, it could make sense to change the signature of `FootballService` to return a null value in case it doesn't find a player. However, if the controller returns a null the response will be HTTP 200 anyway. To avoid this behavior, we added the **@ExceptionHandler** annotation to add a handler method to manage a specific type of exception, and the **@ResponseStatus** annotation to manage the HTTP status code to return in that specific method handler.

There's more...

It is possible to control the response codes more explicitly in your code. Instead of using your data model directly in the controller, you can return `ResponseEntity`, which allows you to specify the status code explicitly. The following is an example of how you can implement `getPlayer` in this way:

```
@GetMapping("/{id}")
public ResponseEntity<Player> readPlayer(@PathVariable String id)
{
  try {
        Player player = footballService.getPlayer(id);
        return new ResponseEntity<>(player, HttpStatus.OK);
    } catch (NotFoundException e) {
        return new ResponseEntity<>(HttpStatus.NOT_FOUND);
    }
}
```

Another alternative is having a global handler for all controllers by using a class annotated with @ ControllerAdvice, like so:

```
package com.packt.football;
@ControllerAdvice
public class GlobalExceptionHandler {
    @ExceptionHandler(NotFoundException.class)
     public ResponseEntity<String>
handleGlobalException(NotFoundException ex) {
         return new ResponseEntity<String>(ex.getMessage(), HttpStatus.
NOT_FOUND);
     }
}
```

In this way, you can have consistent error-handling for all your RESTful endpoints in your application.

Testing a RESTful API

Testing the application manually can be tiring, especially when dealing with challenging scenarios that are hard to validate. Additionally, it lacks scalability in terms of development productivity. Hence, I highly recommend applying automated testing.

By default, Spring Boot includes the *Testing starter* that provides the basic components for unit and integration testing. In this recipe, we'll learn how to implement a unit test for our RESTful API.

Getting ready

In this recipe, we'll create unit tests for the RESTful API created in the previous recipe. I prepared a working version in case you haven't completed it yet. You can find it in the book's GitHub repository at https://github.com/PacktPublishing/Spring-Boot-3.0-Cookbook/. You can find the code to start this recipe in the chapter1/recipe1-4/start folder.

How to do it...

Let's add some tests to our RESTful API that will validate our application whenever we change it:

1. We'll start by creating a new test class for our RESTful controller in the src/test folder. Let's name the new class PlayerControllerTest and annotate it with @WebMvcTest as follows:

    ```
    @WebMvcTest(value = PlayerController.class)
    public class PlayerControllerTest {
    }
    ```

2. Now, define a field of type `MockMvc` and annotate it as `@Autowired`:

```
@Autowired
private MockMvc mvc;
```

3. Then, create another field of type `FootballService` and annotate it with `@MockBean`.

4. We are ready now to write our first test:

I. Let's create a method to validate when our RESTful API returns the players. Name the new method `testListPlayers`:

```
@Test
public void testListPlayers() throws Exception {
}
```

Importantly, note that it should get annotated with `@Test`.

II. The first thing to do in the test is configure `FootballService`. The following lines configure `FootballService` to return a list of two players when invoking the `listPlayers` method:

```
Player player1 = new Player("1884823", 5, "Ivana ANDRES",
"Defender", LocalDate.of(1994, 07, 13));
Player player2 = new Player("325636", 11, "Alexia PUTELLAS",
"Midfielder", LocalDate.of(1994, 02, 04));
List<Player> players = List.of(player1, player2);
given(footballService.listPlayers()).willReturn(players);
```

III. Next, we'll use the `mvc` field created in *step 2* to emulate the HTTP calls and validate it's behaving as expected:

```
MvcResult result = mvc.perform(MockMvcRequestBuilders
                .get("/players")
                .accept(MediaType.APPLICATION_JSON))
                .andExpect(status().isOk())
                .andExpect(MockMvcResultMatchers
                    .jsonPath("$", hasSize(2)))
                .andReturn();
```

The preceding code performs a GET request that accepts application/JSON content. The expected result is OK, meaning any HTTP status code between 200 and 299. The expected result is a JSON array with two elements. Finally, we save the result in the `result` variable.

IV. As we saved the result in the `result` variable, we can perform additional validations. For instance, we can validate that the returned array of players is exactly as expected:

```
String json = result.getResponse().getContentAsString();
ObjectMapper mapper = new ObjectMapper();
mapper.registerModule(new JavaTimeModule());
```

```
List<Player> returnedPlayers = mapper.readValue(json,
                  mapper.getTypeFactory().
constructCollectionType(List.class, Player.class));
assertArrayEquals(players.toArray(), returnedPlayers.toArray());
```

5. Now, we can test that the application manages the errors as expected. Let's test what happens when we request a player that does not exist:

  I. Create a new method named `testReadPlayer_doesnt_exist` in the `PlayerControllerTest` class. Remember to annotate it with @Test:

```
@Test
public void testReadPlayer_doesnt_exist() throws Exception {
}
```

  II. Let's arrange the `getPlayer` method of the `FootballService` class to throw a `NotFoundException` when trying to get the player `1884823`. For that, use the following code:

```
String id = "1884823";
given(footballService.getPlayer(id))
        .willThrow(new NotFoundException("Player not found"));
```

  III. Now we can use the `mvc` field defined in *step 2* to simulate the request and then validate it behaves as expected:

```
mvc.perform(MockMvcRequestBuilders.get("/players/" + id).
accept(MediaType.APPLICATION_JSON))
        .andExpect(status().isNotFound());
```

6. To execute the tests, use the following command:

```
mvn test
```

The `test` goal is also executed anytime you execute the `package` or `install` goals unless you explicitly disable the test execution.

Usually, you can also execute the tests from your favorite IDE.

How it works...

By default, Spring Initializr includes a dependency for `spring-boot-starter-test`. This dependency provides all the necessary components to create tests. Let's describe the elements used in this recipe:

- @WebMvcTest: When you apply this annotation to a testing class, it disables Spring Boot default autoconfiguration and applies only the relevant configuration for MVC tests. That means that it doesn't register classes annotated with @Service as FootballService, but it registers classes annotated with @RestController as PlayerController.

- @MockBean: As our FootballService class is not autoconfigured because we use @WebMvcTest), we can register our own implementation of FootballService. The @MockBean annotation allows us to mock the implementation of FootballService, replacing any previous bean registration.

- given: This method stubs a method and allows us to specify the behavior. For instance, using thenReturn sets the return value when the method specified by given is called.

- MockMvc: This simulates the behavior of the web server and allows you to test your controllers without having to deploy the application. When performing simulated requests, it returns ResultActions that provides methods to validate that the controller behaves as expected, for instance, the andExpect method.

In this recipe, we used other JUnit utilities, such as assertArrayEquals to compare the elements of two arrays. The utilities offered by JUnit and other testing libraries are very extensive and we won't cover them all in detail in this book. However, I'll explain the testing utilities the first time I introduce them as we go through.

There's more...

You can write tests for the rest of the methods exposed by our RESTful API as an exercise. I also prepared some tests for this RESTful API myself. You can find them in the book's GitHub repository at https://github.com/PacktPublishing/Spring-Boot-3.0-Cookbook/ – the final version is in the chapter1/recipe1-4/end folder.

See also

In this book, I apply the **Arrange-Act-Assert (AAA)** principles when writing tests:

- **Arrange**: In this step, you prepare the class you want to test by setting up the conditions needed for the "act" step to run

- **Act**: In this step, you perform the action you are testing

- **Assert**: In this step, you verify that the expected results were achieved

There is also the **Arrange-Act-Assert-Clean (AAAC)** variant that adds a last step to clean up any change done by the test. Ideally, that last step should not be necessary as we can mock any component or service that handles a state that requires being cleaned up.

Using OpenAPI to document our RESTful API

Now that we have a RESTful API, we can create a consumer application. We could create an application and just perform HTTP requests. That would require that we provide the consumers with the source code of our application and that they understand it. But what if they are developing their application in a different language and they don't know Java and Spring Boot? For this reason, OpenAPI was created. OpenAPI is a standard to document RESTful APIs and can be used to generate client applications. It's widely adopted and is supported by different languages and frameworks. Spring Boot's support for OpenAPI is excellent.

In this recipe, we'll learn how to add OpenAPI support to our RESTful API and consume it using the tools provided by OpenAPI.

Getting ready

In this recipe, we will enhance the RESTful API created in the previous recipe. If you haven't completed the previous recipe, you can find a working version in the book's GitHub repository at https://github.com/PacktPublishing/Spring-Boot-3.0-Cookbook.

You can find the code to start this exercise in the chapter1/recipe1-5/start folder.

> **Note**
>
> OpenAPI 3.0 is the new name of Swagger after it was donated by SmartBear to the OpenAPI Initiative. You will likely find a lot of documentation still using the name *Swagger* when referring to OpenAPI.

How to do it...

Let's document our RESTful API with OpenAPI and start testing from the nice OpenAPI user interface:

1. Open the pom.xml file of RESTful API project and add the SpringDoc OpenAPI Starter WebMVC UI dependency, org.springdoc:springdoc-openapi-starter-webmvc-ui. To add the dependency, insert the following XML into the <dependencies> element:

    ```
    <dependencies>
        <dependency>
            <groupId>org.springdoc</groupId>
            <artifactId>springdoc-openapi-starter-webmvc- ui</
    artifactId>
            <version>2.2.0</version>
        </dependency>
    </dependencies>
    ```

> **Important**
>
> For brevity, I removed the other dependencies from the code snippet, but you should keep all of them in your code.

2. Now you can execute this application and open the following URL in your browser: `http://localhost:8080/v3/api-docs`. It returns the description of your RESTful API in OpenAPI format. You can also open `http://localhost:8080/swagger-ui/index.html` for a nice user interface to interact with your API.

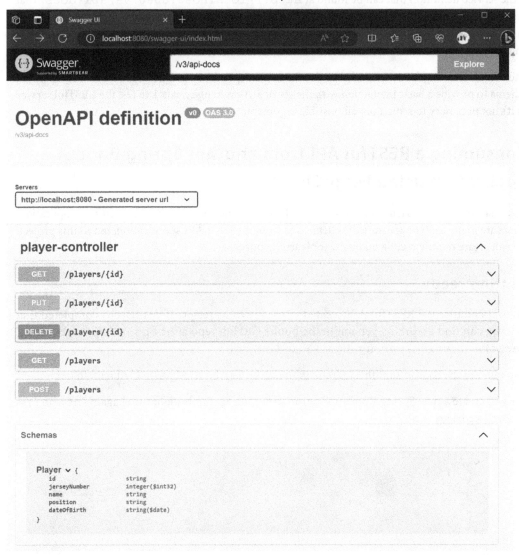

Figure 1.2: Open API (Swagger) UI for our RESTful API

3.   As you can see, it exposes all the RESTful operations defined in the application and the data model used, in this case, `Player`. Now you can use the browser to execute any of the operations available.

How it works...

The `org.springdoc:springdoc-openapi-starter-webmvc-ui` dependency examines the application at runtime to generate the description of the endpoints available. The core of OpenAPI is the service definition that can be found at `http://localhost:8080/v3/api-docs`. That is a JSON document that follows the OpenAPI schema and describes the RESTful endpoints hosted in the application. An endpoint is a combination of a path, HTTP method, parameters, responses, and data schemas.

The other interesting feature provided by the OpenAPI dependency is a nice UI that uses the OpenAPI schema to provide a basic interaction with the service. It can replace `curl` to test the RESTful service as it's not necessary to remember all possible arguments.

Consuming a RESTful API from another Spring Boot application using FeignClient

Now that we have a RESTful API and it's properly documented, we can create a consumer application. There are many tools to generate the client code from the OpenAPI specification, but in this project, we will create the client code manually for learning purposes.

Getting ready

We will enhance the RESTful API created in the previous recipe. If you haven't completed that yet, you can find a working version in the book's GitHub repo at `https://github.com/PacktPublishing/Spring-Boot-3.0-Cookbook`.

You can find the code to start this exercise in the `chapter1/recipe1-6/start` folder.

We will create a new Spring Boot application using the Spring Initializr tool again (`https://start.spring.io`).

How to do it...

We'll create a Spring Boot application consuming the Football RESTful API created in the previous recipe:

1. First, we'll create a new Spring Boot application. Open `https://start.spring.io` and use the same parameters as in the *Creating a RESTful API* recipe, except for changing the following options:

    - For **Artifact**, type `albums`

    - For **Dependencies**, select **Spring Web** and **OpenFeign**

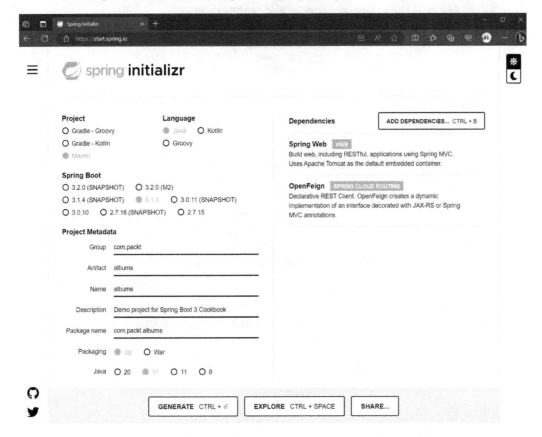

Figure 1.3: Spring Initializr for consumer application

2. Generate the project to download the ZIP file. Extract the project and open `pom.xml`.

3. Create a record named `Player` and add the following code:

    ```
    public record Player(String id, Integer jerseyNumber,
                         String name, String position,
                         LocalDate dateOfBirth) {

    }
    ```

4. Create an interface named `FootballClient` and add the following code:

    ```
    @FeignClient(name = "football", url = "http://localhost:8080")
    public interface FootballClient {
        @RequestMapping(method = RequestMethod.GET,
                        value = "/players")
        List<Player> getPlayers();
    }
    ```

5. Create a controller named `AlbumsController.java` with the following code:

    ```
    @RestController
    @RequestMapping("/albums")
    public class AlbumsController {
        private final FootballClient footballClient;
        public AlbumsController(FootballClient footballClient) {
            this.footballClient = footballClient;
        }
        @GetMapping("/players")
        public List<Player> getPlayers() {
            return footballClient.getPlayers();
        }
    }
    ```

    Modify the `AlbumsApplication` application class by adding the `@EnableFeignClients` annotation:

    ```
    @EnableFeignClients
    @SpringBootApplication
    public class AlbumsApplication {
    }
    ```

6. Now execute the application by executing the following command in your terminal:

    ```
    ./mvnw spring-boot:run \
    -Dspring-boot.run.arguments=-- server.port=8081
    ```

    The additional parameter is to run this application listening on port 8081, instead of the default 8080. The other application is listening on port 8080, hence we need to avoid port conflicts.

7. Then test the application:

    ```
    curl http://localhost:8081/albums/players
    ```

    You will receive a response similar to this:

    ```
    [{"id":"1884823","jerseyNumber":5,"name":"Ivana
    ANDRES","position":"Defender","dateOfBirth":"1994-07-
    13"},{"id":"325636","jerseyNumber":11,"name":"Alexia
    PUTELLAS","position":"Midfielder","dateOfBirth":"1994-02-04"}]
    ```

How it works...

Feign is a **declarative** web service client framework that simplifies making HTTP requests to RESTful web services. You create a Feign client by defining an **interface** that specifies the HTTP requests you want to make to a particular service. The methods in this interface are annotated with annotations similar to @RequestMapping, such as @GetMapping, @PostMapping, and @PutMapping, to specify the HTTP method and the URL path. Indeed, the annotations are the same as those used in the server-side application.

You can inject the Feign client interface into your Spring components and use it to make HTTP requests. Spring Cloud Feign will automatically generate and execute the HTTP requests based on the interface definition. By decorating the application class with @EnableFeignClients, it scans the application for interfaces with the @FeignClient annotation and generates the client.

In the controller, we can then use the Feign client simply via Spring Boot dependency injection.

Do note that we passed an additional parameter, -Dspring-boot.run.arguments=--server. port=8081, to execute the client application. The reason is that the RESTful API is already using port 8080 so we need to execute the client application in a different port.

There's more...

There are other options than Feign to perform the requests. I decided to use Feign due to its great integration with Spring Cloud components, such as Eureka Server. We will see in the following recipes how to integrate with Spring Cloud and how it can do load balancing on the client side.

Most of the code on the client side in this recipe can be automatically generated using IDE integrations or standalone tools. These tools are especially useful to maintain the client-side code in sync with the server descriptions. These tools use OpenAPI descriptions exposed by the RESTful API to generate the client code:

- OpenAPITools: https://github.com/OpenAPITools/openapi-generator
- swagger-codegen: https://github.com/swagger-api/swagger-codegen

Both projects provide a command-line tool and a Maven plugin to generate client-side code.

Consuming a RESTful API from another Spring Boot application using RestClient

In this recipe, we'll use a new component introduced in Spring Framework 6.1 and available in Spring Boot since version 3.2. In the previous recipe, we created a FeignClient by creating an interface in the client application and defining the same methods available in the target service. By using the RestClient component, we will have a fluent API that offers an abstraction over HTTP libraries. It allows converting from Java objects to HTTP requests, and the other way round, the creation of objects from the HTTP responses.

Getting ready

We will enhance the RESTful API created in the *Using OpenAPI to document our RESTful API* recipe. If you haven't completed it yet, you can find a working version in the book's GitHub repo at `https://github.com/PacktPublishing/Spring-Boot-3.0-Cookbook`.

You can find the code to start this exercise in the `chapter1/recipe1-7/start` folder.

We will create a new Spring Boot application using the Spring Initializr tool again (`https://start.spring.io`).

How to do it...

We'll create a new Spring Boot application using the Spring Initializr tool that will consume the RESTful API created in the *Using OpenAPI to document our RESTful API* recipe:

1. Let's start by creating a new Spring Boot application using the Spring Initializr tool. To do this, open `https://start.spring.io` in your browser and use the same parameters as in the *Creating a RESTful API* recipe, except for changing the following options:

    *   For **Artifact**, type `albums`

    *   For **Dependencies**, select **Spring Web**

2. Now, create a configuration class named `AlbumsConfiguration` in which we define a `RestClient` bean:

    ```
    @Configuration
    public class AlbumsConfiguration {
        @Value("${football.api.url:http://localhost:8080}")
        String baseURI;
        @Bean
        RestClient restClient() {
            return RestClient.create(baseURI);
        }
    }
    ```

Note that we defined a field with the @Value annotation to configure the URL of the remote server.

3. Next, create a service class named FootballClientService. This class will use the Spring Boot container to inject the RestClient bean in the constructor:

```
@Service
public class FootballClientService {
    private RestClient restClient;
    public FootballClientService(RestClient restClient) {
        this.restClient = restClient;
    }
}
```

4. Now, you can use the RestClient to retrieve the data from the remote RESTful API. You can create a method named getPlayers as follows:

```
public List<Player> getPlayers() {
    return restClient.get().uri("/players").retrieve()
        .body(new ParameterizedTypeReference<List<Player>>(){ });
}
```

5. Next, you can create another method to get just a single player from the remote RESTful API:

```
public Optional<Player> getPlayer(String id) {
    return restClient.get().uri("/players/{id}", id)
        .exchange((request, response) -> {
            if (response.getStatusCode().equals(HttpStatus.NOT_
FOUND)) {
                return Optional.empty();
            }
            return Optional.of(response.bodyTo(Player.class));
        });
}
```

6. Finally, you can create an Album RESTful API using the FootballClientService service. I created a sample version that you can find in the book's GitHub repository.

How it works...

In this recipe, we didn't create any additional type to replicate the remote RESTful API. Instead, we used the RestClient to perform requests using the Fluent API style, that is, using the result of a method to chain a call to another method. This Fluent API design is easier to read, that's why it's named "Fluent".

Let's analyze what we did in the getPlayer method:

- We started by calling the get method, which returns an object that can be used to set the request's properties, such as the URI, the headers, and other request parameters. We just set the remote address by using the uri method. Note that this address is appended to the base address defined in the AlbumsConfiguration class.

- When we called the exchange method, the RestClient performed the call to the remote RESTful API. Then, the method exchange provides a handler to manage the response.

- In the response handler, we control what happens if the player is not found, in which case we return an empty object. Otherwise, we use the bodyTo method, which allows passing a type to be used to deserialize the response. In this example, we used the Player class.

The code for getPlayers is very similar to getPlayer; the main difference is that the result is a List of players. To specify this, it was necessary to use the ParameterizedTypeReference class to pass a generic type. To capture the generic type, it's necessary to define a subclass of ParameterizedTypeReference, which we did by defining an anonymous inline class. That's why we added the new ParameterizedTypeReference<List<Player>>() { }, including the curly braces { } at the end.

In this recipe, we used the @Value annotation in the AlbumsConfiguration class. This annotation allows us to inject values from external sources, for instance from configuration files or environment variables. The value "${football.api.url:http://localhost:8080}" means that it will try to get the footbal.api.url configuration property first. If it's not defined, it will take the default value, http://localhost:8080.

You will see that the format of the properties will change depending on whether they are defined in the application.properties file or the application.yml file. In the application.properties file, you will see the full property in a single line. That is football.api.url=http://localhost:8080.

On the other hand, the application.yml file can nest the properties, so you will see the following:

```
football:
  api:
    url: http://locahost:8080
```

In this book, I'll use the application.yml file in most of the cases, but you will encounter the application.properties format as well, such as when using the environment variables.

Mocking a RESTful API

The main drawback of using a remote service as we did in the previous recipes is that you need the remote service running when you test your client application. To tackle this scenario, you can **mock** a remote server. By *mock*, I mean simulating the behavior of a component or service, in this case, the remote service.

Mocking a remote dependency in a test can be useful for several reasons. One of the main reasons is that it allows you to test your code in isolation, without having to worry about the behavior of the remote dependency. This can be especially useful if the remote dependency is unreliable or slow, or if you want to test your code in different scenarios that are difficult to reproduce with the remote dependency.

In this recipe, we'll learn how to use Wiremock to mock the remote `Football` service in our Albums application during the testing execution.

Getting ready

In this recipe, we'll create the tests for the application we built in the *Consuming a RESTful API from another Spring Boot application using RestClient* recipe. If you haven't completed that recipe yet, I have prepared a working version of the recipe that can be found in the book's GitHub repository at `https://github.com/PacktPublishing/Spring-Boot-3.0-Cookbook`, in `chapter1/ recipe1-8/start`.

How to do it...

We'll add the Wiremock dependency to our project and then we'll be able to create isolated tests for our Albums application:

1. The first thing to do is to add the Wiremock dependency to the Albums project. To do so, open the `pom.xml` file and add the following dependency:

    ```
    <dependency>
        <groupId>com.github.tomakehurst</groupId>
        <artifactId>wiremock-standalone</artifactId>
        <version>3.0.1</version>
        <scope>test</scope>
    </dependency>
    ```

2. Now, we can create a test class for our `FootballClientService`. Let's name the test class `FootballClientServiceTest`. We'll use the `@SpringBootTest` annotation to pass a property with the remote server address:

    ```
    @SpringBootTest(properties = { "football.api.url=http://
    localhost:7979" })
    public class FootballClientServiceTests {
    }
    ```

3. Then, we need to set up a Wiremock server in the test. Add the following content to the `FootballClientServiceTest` class:

```
private static WireMockServer wireMockServer;
@BeforeAll
static void init() {
    wireMockServer = new WireMockServer(7979);
    wireMockServer.start();
    WireMock.configureFor(7979);
}
```

4. Now, we can declare a `FootballClientService` field that will be injected by Spring Boot. Annotate it with `@Autowired`:

```
@Autowired
FootballClientService footballClientService;
```

5. Then, write a test to validate the `getPlayer` method.

    I.   Name the test `getPlayerTest`:

```
@Test
public void getPlayerTest() {
```

    II. Let's start first by arranging the result of the remote service. Add the following code to the test:

```
WireMock.stubFor(WireMock.get(WireMock.urlEqualTo("/
players/325636"))
            .willReturn(WireMock.aResponse()
            .withHeader("Content-Type", "application/json")
            .withBody("""
                {
                    "id": "325636",
                    "jerseyNumber": 11,
                    "name": "Alexia PUTELLAS",
                    "position": "Midfielder",
                    "dateOfBirth": "1994-02-04"
                }
                """)));
```

    III. Next, call the `getPlayer` method. This method depends on the remote service:

```
Optional<Player> player = footballClientService.
getPlayer("325636");
```

IV.   And then validate the results:

```
Player expectedPlayer =new Player("325636", 11, "Alexia
PUTELLAS", "Midfielder", LocalDate.of(1994, 2, 4));
assertEquals(expectedPlayer, player.get());
```

6.   As an exercise, you can create tests for the rest of methods of the `FootballClientService` class and also other scenarios, such as simulating different responses from the remote server. You can find a few more tests prepared in the book's GitHub repository.

How it works...

Wiremock is a library for API mock testing. It can run as an independent tool or as a library, as we did in this recipe. Wiremock is necessary for tests only, and for that reason, we configured the `scope` dependency as `test`. There is a known incompatibility with Spring Boot version 3.2.x. Spring Boot uses Jetty 12, while Wiremock depends on Jetty 11. To avoid that incompatibility we used the `wiremock-standalone` artifact instead of the `wiremock` artifact, as it includes all required dependencies.

The Wiremock project is not part of the Spring Boot framework, however, it is a popular option for mocking services in Spring projects.

In this recipe, we used the `@SpringBootTest` annotation as it uses the SpringBoot context and allows passing custom environment variables with the `properties` field. We used the properties to pass the address of the remote server where we configured Wiremock. We used a different server address to avoid conflicts with the real remote server in case it was running on the machine for whatever reason.

We also used `@BeforeAll` to run the Wiremock server initialization before each test was executed. In that initialization, we configured the Wiremock server to listen on port `7979`, matching the configuration passed in the `properties` field.

With `StubFor` we configured the desired behavior for the remote server: when receiving a GET request for `/players/325636`, it should return a JSON with the mocked player. The rest is just normal test validation to make sure the result is as expected.

See also

You can find more information about Wiremock on the project web page at `https://www.wiremock.io/`.

2

Securing Spring Boot Applications with OAuth2

Open Authorization 2.0 (**OAuth 2.0**) is an open standard protocol that provides secure authorization for web and mobile applications. It allows users to grant limited access to their resources on one website (called the "resource server") to another website or application (called the "client") without sharing their credentials, such as usernames and passwords. This means that the resource server will never see a user's credentials. OAuth 2.0 is widely used for enabling **single sign-on** (**SSO**), accessing third-party APIs, and implementing secure authorization mechanisms. SSO allows a user to log in to any of several related, yet independent, applications with a single ID. Once logged in to an application, the user is not required to reenter the credentials to access the rest of the applications.

OpenID Connect (**OIDC**) is an open standard for user authentication that's built on top of OAuth 2.0. It's used with OAuth 2.0 to enable secure access to user data. An example of this is when an application allows you to sign in with your Google Account. Usually, they can request access to certain parts of your Google Account profile or permissions to interact with your account on your behalf.

In this chapter, we will learn how to deploy a basic Spring Authorization Server that we'll be using in most of the recipes in this book. Then, we'll learn about the most common scenarios to protect an application, from a RESTful API to a web application. Finally, we'll apply the same concepts but using two popular cloud solutions: Google Accounts for user authentication and Azure AD B2C for an extensible end-to-end authentication experience.

Spring Boot offers great support for OAuth2 and OIDC, regardless of the Identity/Authorization server used. It manages the standard OAuth2/OpenID concepts that are implemented by all vendors.

In this chapter, we will cover the following recipes:

- Setting up Spring Authorization Server
- Protecting a RESTful API using OAuth2
- Protecting a RESTful API using OAuth2 with different scopes
- Configuring an MVC application with OpenID authentication
- Logging in with Google Accounts
- Integrating a RESTful API with a cloud **identity provider** (**IdP**)

Technical requirements

This chapter has the same technical requirements as *Chapter 1*. So, we will need an editor such as Visual Studio Code or IntelliJ, Java OpenJDK 21 or higher, and a tool to perform HTTP requests, such as `curl` or Postman.

For some scenarios, you will need a Redis server. The easiest way to run a Redis server locally is by using Docker.

For the *Logging in with Google Accounts* recipe, you will need a Google Account.

For the *Integrating a RESTful API with a cloud IdP* recipe, I used Azure Entra (formerly known as Azure Active Directory) as an authentication provider. You can create a free account with 200 USD credit at `https://azure.microsoft.com/free/search/`.

All the recipes that will be demonstrated in this chapter can be found at: `https://github.com/PacktPublishing/Spring-Boot-3.0-Cookbook/tree/main/chapter2`

Setting up Spring Authorization Server

Spring Authorization Server is a project under the umbrella of Spring Framework that provides the components you need to create an Authorization Server. In this recipe, you will deploy a very simple Authorization Server that you will use for most of the recipes in this chapter. In the following recipes, you will continue to customize this server to achieve the goals of each exercise.

The configuration of this server is just for demo purposes. The Authorization Server plays a crucial role in managing and granting access to protected resources. If you plan to use it in production, I recommend following the instructions from the project at `https://docs.spring.io/spring-authorization-server/reference/overview.html`. In any case, the principles that we will explain in this book have been adjusted to the OAuth2 specification and well-known practices. For this reason, you will be able to apply what you learn here to any other Authorization Server.

Getting ready

To create the Spring Authorization Server, we will use Spring Initializr. You can open this tool in your browser using `https://start.spring.io/` or use it in your code editor if it's been integrated.

I assume that you have basic knowledge of OAuth2. However, I have added some links in the *See also* section that can be useful if you need to go through some concepts.

How to do it...

In this recipe, we will create a Spring Authorization Server using Spring Initializr and do a very basic configuration to create an application registration. Finally, we'll test the application registration and analyze the results. Follow these steps:

1. Open `https://start.spring.io`, as you did in the *Creating a RESTful API* recipe in *Chapter 1*, and use the same parameters, except change the following options:

    - For **Artifact**, type `footballauth`

    - For **Dependencies**, select **OAuth2 Authorization Server**:

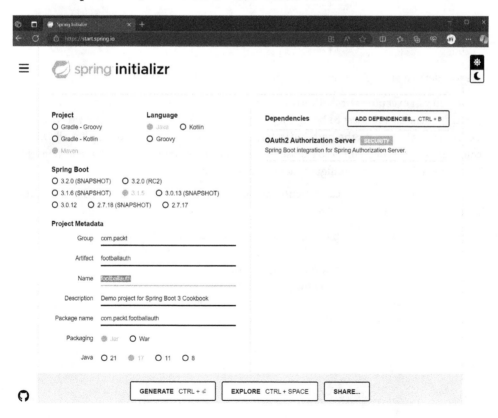

Figure 2.1: Spring Initializr options for Spring Authorization Server

Click on the **GENERATE** button to download the project, then unzip the content to your working folder.

2. Now, we need to configure the authorization server. For that, we will create an `application.yml` file in the `resources` folder with the following content:

```
server:
  port: 9000
spring:
  security:
    oauth2:
      authorizationserver:
        client:
          basic-client:
            registration:
              client-id: "football"
              client-secret: "{noop}SuperSecret"
              client-authentication-methods:
                - "client_secret_post"
              authorization-grant-types:
                - "client_credentials"
              scopes:
                - "football:read"
```

We just defined an application that can be authenticated using the client credential flow.

3. Now, you can execute your Authorization Server. You can retrieve the configuration of our server by making a request to `http://localhost:9000/.well-known/openid-configuration`. As the path indicates, this is a well-known endpoint that all OAuth2-compliant vendors implement to expose the relevant configuration for client applications. Most of the client libraries can configure themselves just from this endpoint.

4. To verify that it works, we can execute the authentication of our client. You can do this by executing the following POST request via `curl`:

```
curl --location 'http://localhost:9000/oauth2/token' \
--header 'Content-Type: application/x-www-form-urlencoded' \
--data-urlencode 'grant_type=client_credentials' \
--data-urlencode 'client_id=football' \
--data-urlencode 'client_secret=SuperSecret' \
--data-urlencode 'scope=football:read'
```

You should see a response that looks similar to this:

```
{"access_token":"eyJraWQiOiIyMWZkYzEyMy05NTZmLTQ5YWQtODU2
Zi1mNjAxNzc4NzAwMmQiLCJhbGciOiJSUzI1NiJ9.eyJzdWIiOiJiYXNp
Yy1jbGllbnQiLCJhdWQiOiJiYXNpYy1jbGllbnQiL
CJuYmYiOjE2OTk1NzIwNjcsInNjb3BlIjpbInByb2ZpbGUiXSwiaXNzIj
oiaHR0cDovL2xvY2FsaG9zdDo5MDAwIiwiZXhwIjoxNjk5NTcyMzY3LCJ
pYXQiOjE2OTk1NzIwNjd9.Tav1nbirP_4zGH8WaJHrcCrNs5ZCnStqqiX
Kc6pakfvQPviosGdgo9vunq4ogRZWYNjXOS5GYwOXlubSj0UDznnxSLyx
7tR7cEZJSQVHc6kffuozycJ_x15yzw6_Kv_pJ4fP00b7pbHWO8ciZKUhmW
-Pvt5TV8sMFY-uNzgsCtiN5EYdplMUfZdwHMy8yon3bUah8Py7RoAw1bIE
ioGUEiK5XLDaE4yGdo8RyyBv4wj3mw6Bs8dcLspLKWXG5spXlZes6XCaSu
0ZXtLE09AgA_Gmq0kwmhWXgnpGKuCkhkXASyJXboQD9TR0y3yTn_aNeiuV
MPzX4DQ7IaCKzgmaYg","scope":"profile","token_type":"Bearer",
"expires_in":299}
```

5. Now, you can copy the value of the `access_token` field, open `https://jwt.ms` in your browser, and paste the value there. In the **Decoded Token** tab, you can see the token in its decoded form, while if you click on the **Claims** tab, you can see an explanation of each field:

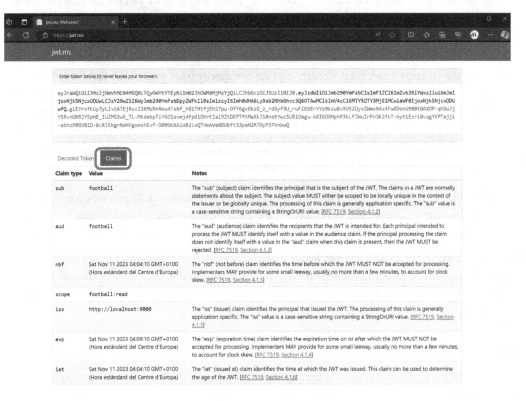

Figure 2.2: JWT token decoded in `jwt.ms`

6. Congratulations – you've deployed a Spring Authorization Server and successfully configured an application for authorization.

How it works...

Spring OAuth2 Authorization Server contains all the components you need to create an authorization server. With the configuration provided in `application.yml`, it created an application with a `client-id` value of `basic-client`. Let's look at the parameters that were used for the application and see how they work:

- The `client-id` is the identifier of the application we create d. In this case, it is `football`.

- The `client-secret` is the secret of the application. By using the `{noop}` prefix in the secret, we tell Spring Security that the password is not encrypted and can be used as-is.

- The `client-authentication-methods` is used to specify how this application can authenticate. By using the `client_secret_post` method, we can ensure that the client ID and secret will be sent in a POST request. We could configure additional methods, such as `client_secret_basic`, in which case the client ID and secret will be sent as HTTP basic schema – that is, in the URL.

- With the `authorization-grant-types`, we specify what grant flows are allowed for this application. By setting `client_credentials`, we are configuring an application that won't have a user interface, such as a background server application. If you have an application that will interact with users, you could configure other options, such as `authorization_code`.

- Finally, with the `scopes`, we are configuring the scopes that are allowed for this application. In this case, it is just the `football:read` scope.

Spring OAuth2 Authorization Server keeps this configuration in memory. As you may have guessed, this is just for demonstration and development purposes. In a production environment, you will need to persist this data. Spring OAuth2 Authorization Server provides support for JPA repositories.

We used **JWT MS** (`https://jwt.ms`) to inspect the access token that was issued by our authorization server. This tool just decodes the token and describes the standard fields. There is another popular tool named **JWT IO** (`https://jwt.io`) that also allows you to validate the token, but it doesn't explain each field.

There's more...

You can follow the instructions from the Spring OAuth2 Authorization Server project to implement the core services with JPA: `https://docs.spring.io/spring-authorization-server/docs/current/reference/html/guides/how-to-jpa.html`.

You can use any relational database supported by Spring Data JPA, such as PostgreSQL, which we used in *Chapter 5*.

See also

In this chapter, we'll manage many OAuth2 concepts, something that can be difficult to understand if you don't have previous knowledge.

For instance, it is very important to understand the different token types:

- `access_token`: This contains all authorization information granted by the authorization server that the resource server will verify.

- `id_token`: This token is used for session management, normally in client applications, to customize the user interface, for example.

- `refresh_token`: This token is used to get new `access_tokens` and `id_tokens` when they are about to expire. `refresh_token` is considered a secret as its lifetime is larger than the others and can be used not only to get fresher tokens for the already authorized applications but also for new ones. It is important to protect this token accordingly.

I strongly recommend getting familiar with the basic OAuth2 flows and their main purposes:

- **Client credential flow**:

    This is the simplest flow and was used in this recipe. It is intended for applications without user interaction – for instance, for server applications communicating with other applications. They can be authenticated in different ways, such as with a secret, as seen in this recipe, a certificate, or other more sophisticated techniques.

- **Authorization code grant flow**:

    This is intended to authenticate web and mobile applications. This is the two-leg authentication flow, where the user authenticates and allows the application to access the requested scopes. Then, the authentication endpoint issues a short-lived piece of code that should be redeemed in the token endpoint to get an access token. After, the application (not the user) should be authenticated. There are two variants of this flow, depending on how it authenticates:

    - Using a client ID and a secret. This is intended for confidential applications, such as those that can keep secrets. This includes server applications.

    - Using a client ID and a challenge, also known as **Proof Key Challenge Exchange** (**PKCE**). This is intended for public applications, such as those that cannot keep a secret, such as mobile applications, or applications that just live in the browser, such as **single-page applications** (**SPAs**).

- **Refresh token flow**:

  As its name suggests, it is used to refresh access and ID tokens when they are about to expire. For that, it uses `refresh_token`.

There are more flows, but these are the basic ones that will be used in this chapter.

Protecting a RESTful API using OAuth2

Protecting a resource – in this case, a RESTful API – is the core functionality of OAuth. In OAuth2, a resource server delegates authorization to access a third-party server – that is, the authorization server. In this recipe, you'll learn how to configure a RESTful API application so that it can authorize the requests that are issued by your Spring Authorization Server.

We will continue with our samples for football data management. You will protect your Football API by only allowing clients who have been granted access by our Authorization Server.

Getting ready

In this recipe, you will reuse the Authorization Server you created in the *Setting up Spring Authorization Server* recipe. If you haven't completed that yet, you can use the authorization server that I've prepared. You can find it in this book's GitHub repository at `https://github.com/PacktPublishing/Spring-Boot-3.0-Cookbook`, in the `chapter4/recipe4-2/start` folder.

How to do it...

In this recipe, you will create a new RESTful API and configure it as a *resource server* using the client registration you created in the previous recipe. Follow these steps:

1. First, create a RESTful API using Spring Initializr (`https://start.spring.io`). Use the same options that you did in the *Creating a RESTful API* recipe in *Chapter 1*, except change the following options:

   - For **Artifact**, type `footballresource`.

   - For **Dependencies**, select **Spring Web** and **Oauth2 Resource Server**:

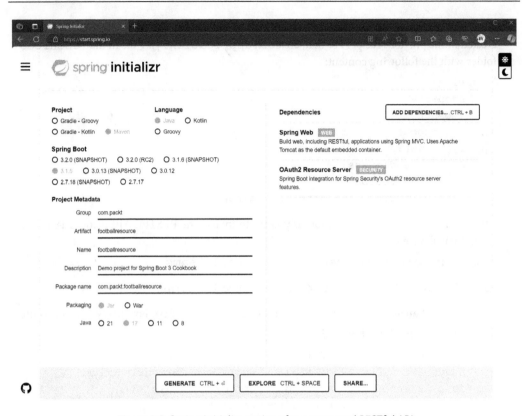

Figure 2.3: Spring Initializr options for a protected RESTful API

Click **GENERATE** to download a ZIP file that contains your project template. Unzip it in your working folder and open it in your code editor.

2. We can create a simple REST controller with a method that returns a list of teams. For that, create a class named `Football.java` with the following content:

```java
@RequestMapping("/football")
@RestController
public class FootballController {
    @GetMapping("/teams")
    public List<String> getTeams() {
        return List.of("Argentina", "Australia",
                       "Brazil");
    }
}
```

Now, let's configure our application for authorization by using the Authorization Server we created in the previous recipe. For that, create an `application.yml` file in the `resources` folder with the following content:

```
spring:
  security:
    oauth2:
      resourceserver:
        jwt:
          audiences:
          - football
          issuer-uri: http://localhost:9000
```

You can now execute the application. If you try to invoke the RESTful API using `curl`, this is what you'll have to do:

```
curl http://localhost:8080/football/teams
```

You will receive an `HTTP Error 401 Unauthorized` error.

3. We need to get an access token from our Authorization Server first. For that, you can execute the following request using `curl`:

```
curl --location 'http://localhost:9000/oauth2/token' \
--header 'Content-Type: application/x-www-form-urlencoded' \
--data-urlencode 'grant_type=client_credentials' \
--data-urlencode 'client_id=football' \
--data-urlencode 'client_secret=SuperSecret' \
--data-urlencode 'scope=football:read'
```

The response will look like this:

Figure 2.4: Access token issued by the Authorization Server

4. Copy the access token value and use it in the next request:

```
curl -H "Authorization: Bearer <access_token>" http://
localhost:8080/football/teams
```

In the preceding request, replace `<access_token>` with the value of the access token you obtained in the previous request, as shown in *Figure 2.4*.

Now, the resource server will return the expected result, along with a list of teams.

5. You now have our RESTful API that's protected by the tokens that were issued by the authorization server.

How it works...

In this simplified example, you saw how authorization works in OAuth2. There's an authorization server issuing a token with authorization information. A resource server – our RESTful API – then checks the token's validity and applies the provided configuration. The authorization server can issue the token in different formats, but the most common is via **JSON Web Tokens (JWT)**. A JWT consists of three parts: a header, a payload, and a signature. These parts are encoded in base64 and are separated by period (.) signs:

- The header contains the metadata needed to manage the token. It is encoded in base64.

- The payload contains the actual data and claims about the token. It carries information such as the expiration time, issuer, and custom claims. It is encoded in base64.

- The signature is created by taking the encoded header and encoded payload, along with a secret, and signing them using the signing algorithm specified in the header. The signature is used to verify the *authenticity* and *integrity* of the token so that we can ensure the token was issued by the authorization server and was not modified by anyone else.

The resource needs to validate the authenticity and integrity of the token. For that, it needs to verify the signature. The Authorization Server provides an endpoint to download the public key of the certificate being used to sign the token. So, the first thing that the authorization server needs to know is where that endpoint is. We can configure this manually in the `application.yml` file, but luckily, Spring Resource Server knows how to retrieve all the information about the authorization server automatically. Just by configuring the `issuer-uri` property, it knows how to retrieve the rest of the information.

Almost all authorization servers in the market can provide the well-known `OpenId` endpoint if we add the following path to the issuer URI: `.well-known/openid-configuration`. The first time the resource server needs to validate a JWT, it calls that endpoint – in our case, `http://localhost:9000/.well-known/openid-configuration` – and retrieves all the information it needs, such as the authorization and token endpoint, the **JSON Web Key Set (JWKS)** endpoint, which contains the sign-in key, and so on. JWKSs are the public keys that the authorization server can use to sign a token. The clients can download these keys to validate the signature of the JWT.

Now that we know how the resource server validates that the token has been issued by the authorization server, we need to know how we can validate that the token is intended for our RESTful APIs. In the `application.yml` file, we've configured the `audiences` field. This indicates the entity for which the token is valid and who or what the token is intended for. The `aud` claim helps ensure that a JWT is only accepted by the intended recipient or resource server. The `aud` claim is part of the payload of the JWT. In our case, the payload, after decoding the base64, looks like this:

```
{
    "sub": "football",
    "aud": "football",
    "nbf": 1699671850,
    "scope": [
      "football:read"
    ],
    "iss": "http://localhost:9000",
    "exp": 1699672150,
    "iat": 1699671850
}
```

Just by setting `issuer-uri` and `audiences`, we ensure that only JWT issued by our Authorization Server and those that are intended for our application/audience will be accepted. Spring Resource Server performs other standard checks, such as for the expiration time (the `exp` claim) and not valid before (the `nbf` claim). Anything else will be rejected with an `HTTP 401 Unauthorized` error. In the *Protecting a RESTful API using OAuth2 with different scopes* recipe, we'll learn how to use other claims to enhance protection.

It is important to note that from a Spring Resource Server perspective, it's not important how the client obtained the access token as that responsibility is delegated to the Authorization Server. The Authorization Server may require different levels of validations, depending on the kind of resource being accessed. Some examples of validations are as follows:

- Client ID and secret, as shown in this example.

- Multiple factors of authentication. For applications with user interaction, the authorization server may consider that the username and password are not enough and force using a second factor of authentication, such as an authentication application, a certificate, and so on.

- If an application tries to access specific scopes, it may require explicit consent. We see this often with social networks when a third-party application needs to access certain parts of our profile or tries to perform special actions, such as publishing on our behalf.

Protecting a RESTful API using OAuth2 with different scopes

In the previous recipe, we learned how to protect our application. In this recipe, we'll learn how to apply more fine-grained security. We need to apply different levels of access to the application: one general form of read access for the consumers of our RESTful API and administrative access so that we can make changes to the data.

To apply different levels of access to the API, we'll use the standard OAuth2 concept of *scopes*. In OAuth 2.0, `scope` is a parameter that's used to specify the level of access and permissions that a client application is requesting from the user and the authorization server. It defines what actions or resources the client application is allowed to perform on behalf of the user. Scopes help ensure that users have control over which parts of their data and resources they grant access to, and they allow for fine-grained access control. In applications with user interaction, granting a scope may imply explicit consent from the user. For applications with no user interaction, it can be configured with administrative consent.

In our football application, you will create two access levels: one for read-only access and another for administrative access.

Getting ready

In this recipe, we'll reuse the Authentication Server from the *Setting up Spring Authorization Server* recipe and the resource server we created in the *Protecting a RESTful API using OAuth2* recipe. If you haven't completed these recipes yet, you can find a working version in this book's GitHub repository at `https://github.com/PacktPublishing/Spring-Boot-3.0-Cookbook`, in the `chapter4/recipe4-3/start` folder.

How to do it...

Let's create the `football:read` and `football:admin` scopes in the Authorization Server and apply the configuration for managing them in the resource server:

1. The first thing you should do is ensure that the scopes are defined in the authorization server. For that, go to the `application.yml` file in the `resources` folder of the project you created in the *Setting up Spring Authorization Server* recipe. If you're using the implementation I provided, as explained in the *Getting ready* section of that recipe, you can find the project in the `footballauth` folder. Ensure that the application mentions the `football:read` and `football:admin` scopes. The application configuration in the `application.yml` file should look like this:

    ```
    spring:
      security:
    ```

```
oauth2:
  authorizationserver:
    client:
      football:
        registration:
          client-id: "football"
          client-secret: "{noop}SuperSecret"
          client-authentication-methods:
            - "client_secret_post"
          authorization-grant-types:
            - "client_credentials"
          scopes:
            - "football:read"
            - "football:admin"
```

2. Let's create an action that requires administrative access to the RESTful API. For instance, you can create a method to create a team. For that, in the `FootballController` controller class, create a method named `addTeam` that's mapped to a POST action:

```
@PostMapping("/teams")
public String addTeam(@RequestBody String teamName){
  return teamName + " added";
}
```

You can do a more complex implementation, but for this exercise, we can keep this emulated implementation.

3. Now, configure the resource server so that it can manage the scopes. For that, create a configuration class that exposes a `SecurityFilterChain` bean:

```
@Configuration
public class SecurityConfig {
    @Bean
    public SecurityFilterChain
    filterChain(HttpSecurity http) throws Exception {
        return http.authorizeHttpRequests(authorize ->
            authorize.requestMatchers(HttpMethod.GET,
            "/football/teams/**").hasAuthority(
            "SCOPE_football:read").requestMatchers(
            HttpMethod.POST, "/football/teams/**")
            .hasAuthority("SCOPE_football:admin")
            .anyRequest().authenticated())
            .oauth2ResourceServer(oauth2 ->
            oauth2.jwt(Customizer.withDefaults()))
            .build();
```

```
        }
    }
```

Note that as part of `SecurityFilterChain`, we defined a couple of `requestMatchers` with `HttpMethod`, the request path, and the required authority using both scopes.

4. Now that we have the required configuration, let's run the application and perform some tests to validate its behavior:

    I. First, get an access token from the authorization server, requesting just the `football:read` scope. You can execute the request by running the following `curl` command:

    ```
    curl --location 'http://localhost:9000/oauth2/token' \
    --header 'Content-Type: application/x-www-form-urlencoded' \
    --data-urlencode 'grant_type=client_credentials' \
    --data-urlencode 'client_id=football' \
    --data-urlencode 'client_secret=SuperSecret' \
    --data-urlencode 'scope=football:read'
    ```

    II. Take the access token that was returned and use it in the authorization header of the following request:

    ```
    curl -H "Authorization: Bearer <access_token>" http://
    localhost:8080/football/teams
    ```

It should return the data normally and the HTTP response code should be 200.

However, let's say you try to perform the POST request to create a team:

```
curl -H "Authorization: Bearer <access_token>" -H "Content-
Type: application/text" --request POST --data 'Senegal' http://
localhost:8080/football/teams -v
```

A note on curl

Note that we're starting to use the `-v` parameter. It provides a verbose response so that we can see the reasons something fails.

It will return an HTTP 403 forbidden error, and the details will look as this:

```
WWW-Authenticate: Bearer error="insufficient_scope", error_
description="The request requires higher privileges than
provided by the access token.", error_uri="https://tools.ietf.
org/html/rfc6750#section-3.1"
```

III.   Now, let's retrieve another access token with the appropriate scopes:

```
curl --location 'http://localhost:9000/oauth2/token' \
--header 'Content-Type: application/x-www-form-urlencoded' \
--data-urlencode 'grant_type=client_credentials' \
--data-urlencode 'client_id=football' \
--data-urlencode 'client_secret=SuperSecret' \
--data-urlencode 'scope=football:read football:admin'
```

Note that we can request more than one scope at a time.

If we execute the request to create a team with the new access token, we'll see that it works as expected and will return something like `Senegal added`.

5.   With that, our application is protected and we've applied different levels of protection to our resource server.

How it works...

The `SecurityFilterChain` bean is a component that's used for configuring the security filters that intercept and process incoming HTTP requests. Here, we created a `SecurityFilterChain` bean that looks for two matching patterns: GET requests that match the `/football/teams/**` path pattern and POST requests that match the same path pattern. GET requests should have the `SCOPE_football:read` authority and POST should have the `SCOPE_football:admin` authority. Once you configure `SecurityFilterChain`, it is applied to all incoming HTTP requests. Then, if a request matching the pattern doesn't have the required scope, it will raise an `HTTP 403 forbidden` response.

Why is `SCOPE_ prefix` used? It is created by the default `JwtAuthenticationConverter`. This component transforms the JWT into an `Authentication` object. The default `JwtAuthenticationConverter` is wired by Spring Security, but you can also register your own converter if you want a different behavior.

There's more...

More validations can be performed on JWT. For instance, a common way of validating a request is by checking its roles.

You can validate a request's roles by registering a `SecurityFilterChain` bean. Let's say you have an administrator role defined in your Authorization Server. Here, you can configure a `SecurityFilterChain` bean in the resource server to ensure that only users with the ADMIN role can perform POST requests on `football/teams` path, as follows:

```
@Bean
public SecurityFilterChain filterChainRoles(HttpSecurity
http) throws Exception {
```

```
    return http.authorizeHttpRequests(authorize ->
    authorize.requestMatchers(HttpMethod.POST,
    "football/teams/**").hasRole("ADMIN")
    .anyRequest().authenticated())
    .oauth2ResourceServer(oauth2 ->
    oauth2.jwt(Customizer.withDefaults()))
    .build();
}
```

You can make other checks to validate your token. In this case, you can use OAuth2TokenValidator. For instance, you may want to validate that a given claim is present in your JWT. For that, you can create a class that implements OAuth2TokenValidator:

```
class CustomClaimValidator implements
OAuth2TokenValidator<Jwt> {
    OAuth2Error error = new OAuth2Error("custom_code",
    "This feature is only for special football fans",
    null);
    @Override
    public OAuth2TokenValidatorResult validate(Jwt jwt) {
        if (jwt.getClaims().containsKey("specialFan")){
            return OAuth2TokenValidatorResult.success();
        }
        else{
            return
                OAuth2TokenValidatorResult.failure(error);
        }
    }
}
```

See also

I recommend that you look at the Spring OAuth2 Resource Server project documentation at https://docs.spring.io/spring-security/reference/servlet/oauth2/resource-server/jwt.html for more details.

Configuring an MVC application with OpenID authentication

We want to create a new web application for our football fans. To do so, we must authenticate the users when they're accessing the application. We'll use access tokens to access the protected RESTful API.

In this recipe, we'll learn how to use Spring OAuth2 Client to protect an MVC web application and get access tokens for other protected resources.

If you plan to use an SPA, you will need to look for OpenID-certified libraries for your target environment.

Getting ready

For this recipe, you will reuse the Authorization Server application you created in the *Setting up Spring Authorization Server* recipe and the application you created in the *Protecting a RESTful API using OAuth2 with different scopes* recipe. I've prepared a working version of both projects in case you haven't completed them yet. You can find them in this book's GitHub repository at `https://github.com/PacktPublishing/Spring-Boot-3.0-Cookbook`, in the `chapter4/recipe4-4/start` folder.

The authentication process involves some redirections and requires managing sessions for the protected application. We'll use Redis to maintain the sessions for the application. You can download Redis and execute it on your computer, but as we did for other recipes, you can deploy Redis on Docker. For that, just execute the following command in your terminal:

```
docker run --name spring-cache -p 6379:6379 -d redis
```

This command will download the Redis community image if it is not yet present on your computer and will start the Redis server so that it's listening on port 6379 without any credentials. In a production environment, you probably want to secure this service, but in this recipe, we'll keep it open for simplicity.

How to do it...

In this recipe, you'll create a new web application and integrate it with the existing authorization server and the RESTful API you created in previous recipes. Follow these steps:

1. First, you'll need to create the client registration in the Authorization Server. For that, open the `application.yml` file in the `resources` folder of the Authorization Server – that is, the project you created in the *Setting up Spring Authorization Server* recipe. Add the new client registration, as follows:

    ```
    spring:
      security:
        oauth2:
          authorizationserver:
            client:
              football-ui:
                registration:
                  client-id: "football-ui"
                  client-secret: "{noop}TheSecretSauce"
    ```

```
                    client-authentication-methods:
                      - "client_secret_basic"
                    authorization-grant-types:
                      - "authorization_code"
                      - "refresh_token"
                      - "client_credentials"
                    redirect-uris:
                      - "http://localhost:9080/login/oauth2/code/
      football-ui"
                    scopes:
                      - "openid"
                      - "profile"
                      - "football:read"
                      - "football:admin"
                  require-authorization-consent: true
```

2. As we want to authenticate users, you will need to create at least one user. To do that, in the same `application.yml` file in the Authorization Server, add the following configuration:

```
User:
    name: "user"
    password: "password"
```

The `user` element should be aligned with the `oauth2` element. Remember that the indentation in `.yml` files is very important. You can change the username and password and set the values you wish.

Keep this configuration as you'll use it later in the web application.

3. Now, let's create a new Spring Boot application for our web application. You can use *Spring Initializr*, as you did in the *Creating a RESTful API* recipe in *Chapter 1*, but change the following options:

*   For **Artifact**, type `footballui`

*   For **Dependencies**, select **Spring Web, Thymeleaf, Spring Session, Spring Data Redis (Access+Driver), OAuth2 Client**, and **OAuth2 Security**:

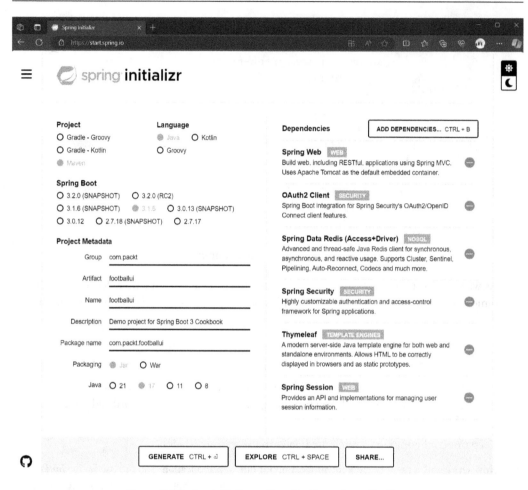

Figure 2.5: Spring Initializr options for the web application

Click **GENERATE** to download the project template as a ZIP file. Unzip the file in your development folder and open it in your preferred code editor.

4. There is a known incompatibility between **Thymeleaf** and **Spring Security**. To integrate both components, it is necessary to add an additional dependency to org.thymeleaf.extras: thymeleaf-extras-springsecurity6. For that, open the project's pom.xml file and add the following dependency:

```
<dependency>
    <groupId>org.thymeleaf.extras</groupId>
    <artifactId>thymeleaf-extras-springsecurity6
        </artifactId>
    <version>3.1.1.RELEASE</version>
</dependency>
```

5. Let's start by creating our web pages. Since we're following the **Model View Controller** (**MVC**) approach, we'll need to create the `Controller` class, which will populate a model and then be presented in the view. The view is rendered using the Thymeleaf template engine:

    I. First, create the `Controller` class and name it `FootballController` while providing a simple method for the home page:

    ```
    @Controller
    public class FootballController {
        @GetMapping("/")
        public String home() {
            return "home";
        }
    }
    ```

    II. This method returns the name of the view. Now, we need to create the view.

    III. For the view, we should create a new template for Thymeleaf. The default template location is the `resources/template` folder. In the same folder, create a file named home . html with the following content:

    ```
    <!DOCTYPE HTML>
    <html>
    <head>
        <title>The great football app</title>
        <meta http-equiv="Content-Type"
            content="text/html; charset=UTF-8" />
    </head>
    <body>
        <p>Let's see <a href="/myself"> who you are </a>.
        You will need to login first!</p>
    </body>
    </html>
    ```

    This is a very basic page, but it now contains a link that we want to protect.

6. Now, we must configure the application so that it can be authenticated using the Authorization Server and force it to be authenticated for all pages except for the home page we just created:

    I. To integrate the web application with the Authorization Server, open the `application.yml` file in the `resources` folder and configure the OAuth2 client application, as follows:

    ```
    spring:
      security:
        oauth2:
          client:
            registration:
    ```

```
        football-ui:
          client-id: "football-ui"
          client-secret: "TheSecretSauce"
          redirect-uri:
            "{baseUrl}/login/oauth2/code/
            {registrationId}"
          authorization-grant-type:
            authorization_code
          scope:
            openid,profile,football:read,
            football:admin
      provider:
        football-ui:
          issuer-uri: http://localhost:9000
```

II.   You'll need the parameters that you defined in the Authorization Server.

III.   Now, configure the application so that it protects all pages except the home page and uses the OAuth2 login credentials. For that, create a new configuration class named `SecurityConfiguration` and create a `SecurityFilterChain` bean:

```
@Configuration
@EnableWebSecurity
public class SecurityConfiguration {
    @Bean
    public SecurityFilterChain
    defaultSecurityFilterChain(HttpSecurity http)
    throws Exception {

        http.authorizeHttpRequests(
            (authorize) -> authorize
            .requestMatchers("/").permitAll()
            .anyRequest().authenticated())
            .oauth2Login(Customizer.withDefaults());
        return http.build();
    }
}
```

With this configuration, you've protected all pages, except the root. So, let's create the rest of the pages.

7.   Next, we need to create a page to show the user information. To do that, in the same `FootballController` controller, create a new method, as follows:

```
@GetMapping("/myself")
public String user(Model model,
@AuthenticationPrincipal OidcUser oidcUser) {
```

```
    model.addAttribute("userName",
        oidcUser.getName());
    model.addAttribute("audience",
        oidcUser.getAudience());
    model.addAttribute("expiresAt",
        oidcUser.getExpiresAt());
    model.addAttribute("claims",
        oidcUser.getClaims());
    return "myself";
}
```

Here, we are asking Spring Boot to inject `OidcUser` as a method parameter and we are creating a model that we'll use in the view named `myself`.

Now, create a file named `myself.html` in the `resources/templates` folder. Put the following content in `<body>` to show the `Model` data:

```
<body>
    <h1>This is what we can see in your OpenId
        data</h1>
    <p>Your username <span style="font-weight:bold"
        th:text="${userName}" />! </p>
    <p>Audience <span style="font-weight:bold"
        th:text="${audience}" />.</p>
    <p>Expires at <span style="font-weight:bold"
        th:text="${expiresAt}" />.</p>
    </div>
    <h2>Here all the claims</h2>
    <table>
        <tr th:each="claim: ${claims}">
            <td th:text="${claim.key}" />
            <td th:text="${claim.value}" />
        </tr>
    </table>
    <h2>Let's try to use your rights</h2>
    <a href="/teams">Teams</a>
</body>
```

As you can see, there is a link to `/teams`. This link will open a new page showing the teams. The teams page retrieves data from the RESTful API you created in the *Protecting a RESTful API using OAuth2 with different scopes* recipe.

8. Let's create a new method in `FootballController` so that we can get the teams. For that, you will get an access token by using the Spring Boot OAuth2 client:

```
@GetMapping("/teams")
public String teams(@
RegisteredOAuth2AuthorizedClient("football-ui")
```

```
OAuth2AuthorizedClient authorizedClient, Model model) {
    RestTemplate restTemplate = new RestTemplate();
    HttpHeaders headers = new HttpHeaders();
    headers.add(HttpHeaders.AUTHORIZATION,
        "Bearer " + authorizedClient.getAccessToken()
        .getTokenValue());
    HttpEntity<String> entity = new HttpEntity<>(null,
        headers);
    ResponseEntity<String> response =
        restTemplate.exchange(
        "http://localhost:8080/football/teams",
        HttpMethod.GET, entity, String.class);
    model.addAttribute("teams", response.getBody());
    return "teams";
}
```

You need to pass the access token in the `Authorization` header, with the `Bearer` string as a prefix.

9. To allow the web application to use the RESTful API, you'll need to include `football-ui`, the web application audience, as an accepted audience. For that, in the project you created in the *Protecting a RESTful API using OAuth2 with different scopes* recipe, open the `application.yml` file in the `resources` folder and add `football-ui` to the `audiences` property. The `application.yml` file should look like this:

```
spring:
  security:
    oauth2:
      resourceserver:
        jwt:
          audiences:
          - football
          - football-ui
          issuer-uri: http://localhost:9000
```

10. There is still an important detail we must cover before we start the new application: we need to configure Redis. The only settings that are required for it are `hostname` and `port`. For that, open the `application.yml` file of the `football-ui` project again and set the following configuration:

```
spring:
  data:
    redis:
      host: localhost
      port: 6379
```

11. The last setting to configure is the application port. The resource server is already using port 8080. To avoid port conflicts, we need to change the port of the football-ui project. For that, in the same application.yml file, add the following setting:

```
server:
  port: 9080
```

You can set any port that is not being used yet. Keep in mind that it's part of the configuration in the Authorization Server. If you modify the port, you will need to modify the configuration in the authorization server.

12. Now, you can run the application. Go to http://localhost:9080 in your browser:

Let's see who you are . You will need to login first!

Figure 2.6: The application's home page

Here, you will see the home page, which is the only route that is not protected. If you click on the **who you are** link, you will be redirected to the login page in the Authorization Server:

Figure 2.7: Login page in the authorization server

Apply the username and password you configured in *Step 2* and click on the **Sign in** button. You'll be redirected to the consent page, where you should permit the scopes being requested by the application:

Figure 2.8: Consent page in the authorization server

Click **Submit Consent**; you will be redirected to the application. The OAuth client application will complete the process by redeeming the access code generated by the authentication endpoint to obtain the ID token, access token, and refresh token. This part of the process is transparent for you as it's managed by the OAuth2 client.

Once it's done this, you'll be returned to the application on the page you created to show the user authentication data:

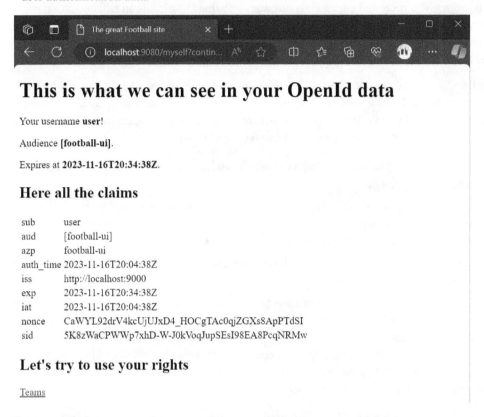

Figure 2.9: Application page with user OpenID data

If you click on the **Teams** link, it will call the RESTful API to get the `teams` data:

Teams: ["Argentina","Australia","Brazil","Canada","China PR","Colombia","Costa Rica","Denmark","England","France","Germany","Italy","Jamaica","Japan","Korea Republic","Morocco","Netherlands","New Zealand","Nigeria","Norway","Panama","Philippines","Portugal","Republic of Ireland","South Africa","Spain","Sweden","Switzerland","USA","Vietnam","Zambia"]

Figure 2.10: The application page showing the data from the RESTful API

With that, your web application is protected by OpenID and it can call another OAuth2-protected resource.

How it works...

In this project, you protected your web application by using OIDC. It is an authentication protocol that is an extension of OAuth 2.0. It provides a standardized way for users to log in to web applications or mobile apps using their existing accounts with IdPs. In our exercise, we used the Authorization Server as an IdP as well.

The OIDC server normally provides a discovery endpoint at `.well-known/openid-configuration`. If this is not provided, then it was likely hidden intentionally by the administrator. That endpoint provides all the information the client applications need to authenticate. In our application, we used the *authorization code grant flow*. It involves several steps:

1. First, the client application redirects the user for authentication, requesting the required scopes for application usage.

2. Then, the user is authenticated. Depending on the features provided by the authorization server, it could use sophisticated mechanisms to validate the user, such as certificates, multiple authentication factors, or even biometric features.

3. If the user is authenticated, the authorization server may ask the user for consent or not, depending on the scopes requested by the client application.

4. Once authenticated, the Authorization Server will redirect the user to the client application, providing a short-lived authorization code. Then, the client application will redeem the authorization code on the token endpoint (provided by the discovery endpoint). The authorization server will return the tokens with the scopes that have been consented to. The scopes that have not been consented by the user won't be present in the issued tokens. The following tokens are returned by the authorization server:

    *   An ID token containing the session information. This token should not be used for authorization, just for authentication purposes.

    *   An access token, which contains the authorization information, such as the scopes that have been consented. If the application requires a scope, it should validate the scopes returned and manage them accordingly.

    *   A refresh token, which is used to get new tokens before they expire.

Since many redirects are involved in this process, the client application needs to keep the state of the user, hence the requirement for session management. The Spring Framework provides a convenient way to manage sessions using Redis.

Keep in mind that the client application needs to access the discovery endpoint when it starts. For that reason, remember to start your authorization server before the client application.

In this exercise, you configured the root page as the only permitted page without authentication. To access any other page, it must be authenticated. For that reason, just by trying to navigate to /myself or /teams, the authorization process is initiated.

See also

Many modern applications are SPA. This type of application runs mostly in the browser. Also, note that many libraries implement OIDC. I recommend using an OpenID-certified library as they are validated by peers.

Even with the growing popularity of SPA, I haven't explained the authentication of this type of application as it is not related to Spring Boot. However, if you are interested in integrating an SPA with Spring Authorization Server, I recommend following the guidelines from the Spring OAuth2 Authorization Server project at https://docs.spring.io/spring-authorization-server/docs/current/reference/html/guides/how-to-pkce.html.

Logging in with Google Accounts

You have a new requirement for your football application: your users want to log in to your application using their Gmail accounts.

To implement this scenario, you will configure your Authorization Server as an OAuth2 client, with Google Accounts being their IdP. You will learn how to create an OAuth2 Client ID in Google Cloud and integrate it into your application.

Getting ready

For this recipe, you will reuse the Spring Authorization Server you created in the *Setting up Spring Authorization Server* and *Protecting a RESTful API using OAuth2 with different scopes* recipes, as well as the web application you created in the *Configuring an MVC application with OpenID authentication* recipe. The MVC application stores sessions in Redis. You can run a Redis server in Docker, as explained in the *Configuring an MVC application with OpenID authentication* recipe.

I've prepared a working version of the required recipes in case you haven't completed them yet. You can find them in this book's GitHub repository at https://github.com/PacktPublishing/Spring-Boot-3.0-Cookbook, in the chapter4/recipe4-5/start folder.

As you will integrate the application with Google Accounts, you will need a Google Account. If you don't have a Google Account yet, you can create one at https://accounts.google.com.

How to do it...

Let's start by creating an OAuth2 client in Google and then use the configuration provided to configure our Authorization Server so that it can log in to your application using GoogleAaccounts:

1. Let's start by opening the Google Cloud console at `https://console.cloud.google.com/`. You will need to log in using your Google Account. Once you've done this, you will see the Google Cloud home page:

    I.   First, you'll need to create a project:

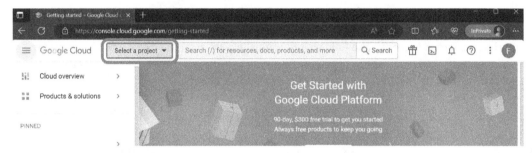

Figure 2.11: The Google Cloud home page

    II.   To create the project, click **Select a project**.

    III. In the **Select a project** dialogue, click **NEW PROJECT**:

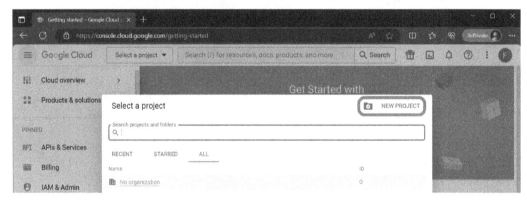

Figure 2.12: Creating a new project

IV.   Name the project – for instance, `springboot3-cookbook`:

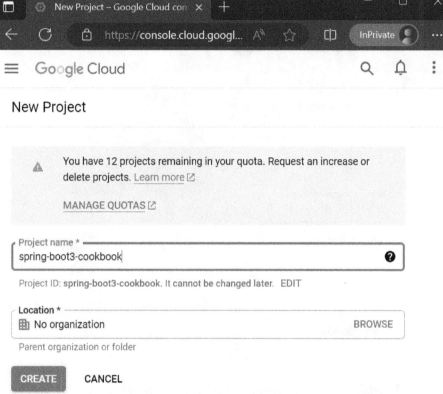

Figure 2.13: New Project settings

V.   Click **CREATE**. This process will take a few moments. A notification will appear when it is completed. Once created, select the project.

2. Now that we have a project, let's configure the consent page for our web application. For that, open the **APIs & Services** menu and choose **Credentials**:

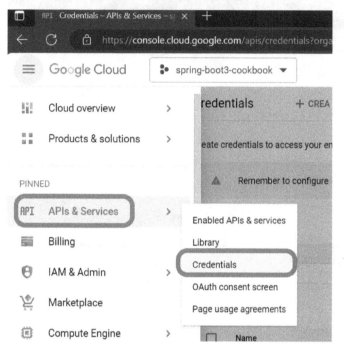

Figure 2.14: The Credentials menu

On the **Credentials** page, you will see a reminder to create the consent page for the application:

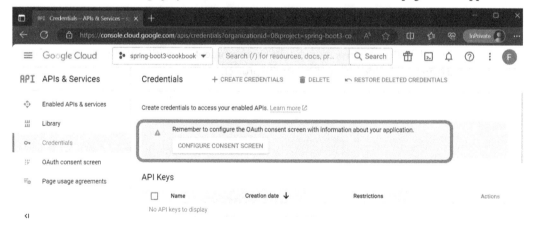

Figure 2.15: The Credentials home page with a reminder to
configure the OAuth consent screen highlighted

Click on the **CONFIGURE CONSENT SCREEN** button and follow the instructions to configure the consent page. For **User Type**, select **External** and click **CREATE**.

With this type of user, the application will start in testing mode. This means that only some test users will be able to use it. Once you've completed the development process and the application is ready, you can publish it. To do this, your site must be verified. We won't publish the application in this recipe, but if you plan to use it in your application, you'll need to complete this step.

After selecting **External** for **User Type**, there is a four-step process you must go through:

I.   First, we have the **OAuth consent** screen:

- Here, you should configure the application's name. You can set it as **Spring Boot 3 Cookbook**, for instance.

- You should configure a user support email. You can use the same email address that you use for your Google Account.

- You should also configure the Developer Contact Information email. Again, you can use the same email address that you use for your Google Account.

- The rest of the parameters are optional. We don't need to configure them for now.

II.   For **Update selected scopes**, click **ADD OR REMOVE SCOPES** and select **openid, userinfo.email**, and **userinfo.profile**:

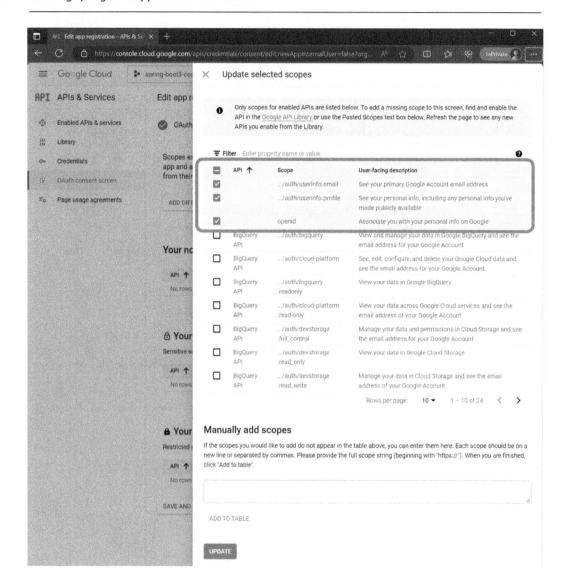

Figure 2.16: Selecting scopes

III.   Then, click **UPDATE**.

IV.   In **test users** step, click **ADD USERS** to add some testing users who will be able to access the application before it's published. You can add a different Google Account to test the application.

V.   In the **summary** step, you will see a summary of your consent screen. You can click **BACK TO DASHBOARD** to return to the **OAuth consent screen** page.

3.   Next, we will create the client credentials. For that, once again, navigate to the **Credentials** page, as shown in *Figure 2.17*. Once you are on the **Credentials** page, click **+ CREATE CREDENTIALS** and select **OAuth client ID**:

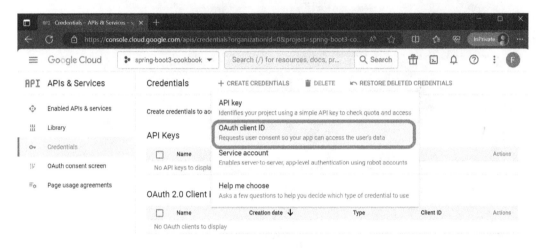

Figure 2.17: Selecting OAuth client ID credentials

- For **Application type**, select **Web application**

- For **Name**, set `football-gmail`

- For **Authorized redirect URIs**, add `http://localhost:9000/login/oauth2/code/football-gmail`:

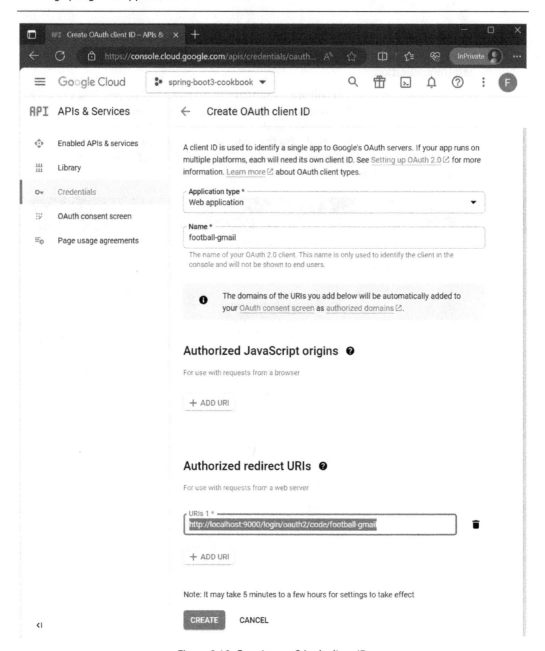

Figure 2.18: Creating an OAuth client ID

Click **CREATE**. A dialogue box will appear, informing you that the client was created and showing you the **Client ID** and **Client secret** details. We'll need this data to configure the Authorization Server, so keep it safe:

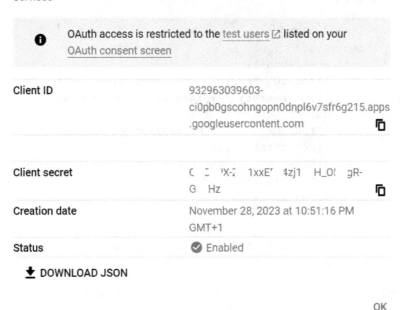

Figure 2.19: OAuth client created

There is a button to download a JSON file containing the configuration. Click on it and save the JSON file in a secure place as it contains the credentials that are required to use the client.

4. Now, we can configure the Spring OAuth2 Authorization Server:

I.   First, we need to add the OAuth2 client dependency. For that, open the pom.xml file and add the following dependency:

```
<dependency>
    <groupId>org.springframework.boot</groupId>
    <artifactId>spring-boot-starter-oauth2-client
        </artifactId>
</dependency>
```

II.   Next, open the `application.yml` file in the `resources` folder and add the client configuration for Google:

```
spring
  security:
    oauth2:
      client:
        registration:
          football-gmail:
            client-id: "replace with your client id"
            client-secret: "replace with your secret"
            redirect-uri:
              "{baseUrl}/login/oauth2/code/
              {registrationId}"
            authorization-grant-type:
              authorization_code
            scope: openid,profile,email
        provider:
          football-gmail:
            issuer-uri: https://accounts.google.com
            user-name-attribute: given_name
```

III. Replace the `client-id` and `client-secret` fields with the values you obtained in *Step 2*.

IV.   The last step to configure the Authorization Server is defining the behavior of the security checks. For that, create a configuration class named `SecurityConfig`:

```
@Configuration
@EnableWebSecurity
public class SecurityConfig {
}
```

V.   Then, add a `SecurityFilterChain` bean:

```
@Bean
@Order(1)
public SecurityFilterChain
authorizationServerSecurityFilterChain(HttpSecurity http)
        throws Exception{
    OAuth2AuthorizationServerConfiguration
        .applyDefaultSecurity(http);
    http.getConfigurer(
        OAuth2AuthorizationServerConfigurer.class)
        .oidc(Customizer.withDefaults());
    http
```

```
        .exceptionHandling((exceptions) -> exceptions
            .defaultAuthenticationEntryPointFor(
                new LoginUrlAuthenticationEntryPoint(
                    "/oauth2/authorization/
                    football-gmail"),
                new MediaTypeRequestMatcher(
                MediaType.TEXT_HTML)
            )
        )
        .oauth2ResourceServer((oauth2) ->
            oauth2.jwt(Customizer.withDefaults()));

    return http.build();
}
```

The preceding code configures Spring Security to use OAuth 2.0 and OpenID Connect 1.0 for authentication, as well as to accept JWT access tokens for certain requests. For instance, it will accept requests that provide JWT access tokens to get user information.

You will also need to add another `SecurityFilterChain` bean but with less priority. It will initiate the OAuth2 login process, which means it will initiate the authentication as a client application in Google, as configured in the `application.yml` file:

```
@Bean
@Order(2)
public SecurityFilterChain
defaultSecurityFilterChain(HttpSecurity http)
        throws Exception {
    http
        .authorizeHttpRequests((authorize) ->
            authorize
            .anyRequest().authenticated())
        .oauth2Login(Customizer.withDefaults());
    return http.build();
}
```

The preceding code configures Spring Security to require authentications for all requests and to use OAuth 2.0 for logging in.

5. With that, your OAuth2 Authorization Server has been configured to require authentication using Google Accounts. Now, you can run all your environments:

    ▪ Run the Authorization Server you just configured

    ▪ Run the RESTful API server you created in the *Protecting a RESTful API using OAuth2 with different scopes* recipe

- Run the web application you created in the *Configuring an MVC application with OpenID authentication* recipe

When you navigate to `http://localhost:9080/myself`, you will be asked to log in using a Google Account:

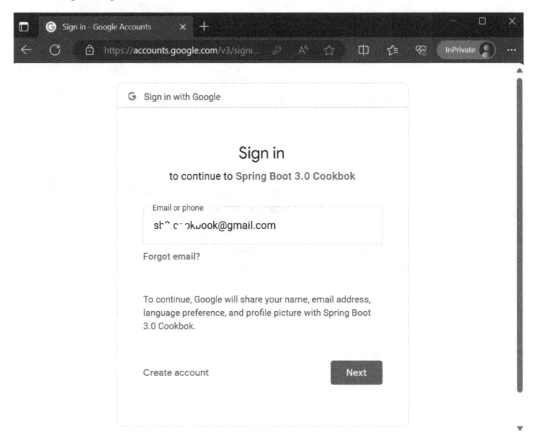

Figure 2.20: Logging in using a Google Account

Once you've logged in, you will see that the application is the same. You can now use the claims issued by the Authorization Server and invoke the RESTful API.

How it works...

Because we're using the Authorization Server with just the login delegated in Google, the Authorization Server doesn't need to maintain a user repository, though it still has the responsibility of issuing tokens. This means that when the application requests to identify a user, the Authorization Server redirects the login to Google. Once back, the Authorization Server continues issuing tokens. Due to this, you didn't need to change the code in the MVC web application or the RESTful API.

It is possible to configure the MVC application so that it bypasses the authorization server and logs in to Google Accounts directly. You just need to replace the OAuth2 client configuration in the MVC web application with the client configuration you used in the Authorization server. However, in that case, you won't be able to use the access tokens that have been issued by Google to protect your RESTful API. This is because Google access tokens are intended to be used with Google services only and they are not standard JWT.

The main complexity is configuring the security chain as there are many options available. In the `SecurityConfig` class, there are two `Beans` with different priorities.

The `SecurityConfig` class defines two `SecurityFilterChain` beans. Here, `SecurityFilterChain` is essentially a chain of filters that Spring Security uses to perform various security checks. Each filter in the chain has a specific role, such as authenticating the user.

The first `SecurityFilterChain` bean is defined with an order of 1, meaning it will be the first filter chain to be consulted. This filter chain has been configured to apply default security settings for an OAuth 2.0 authorization server. It also enables OpenID Connect 1.0 via the `oidc(Customizer.withDefaults())` method call.

The configuration also specifies that if a user is not authenticated, they should be redirected to the OAuth 2.0 login endpoint. This is done using `LoginUrlAuthenticationEntryPoint` with a URL of `/oauth2/authorization/football-gmail`.

The filter chain is also configured to accept JWT access tokens for user info and/or client registration. This is done using the `oauth2ResourceServer((oauth2) -> oauth2.jwt(Customizer.withDefaults()))` method call.

The second `SecurityFilterChain` bean is defined with an order of 2, meaning it will be consulted if the first filter chain does not handle the request. The `anyRequest().authenticated()` chain of methods means that any request must be authenticated.

The `oauth2Login(Customizer.withDefaults())` method call configures the application to use OAuth 2.0 for authentication. The `Customizer.withDefaults()` method call is used to apply the default configuration for OAuth 2.0 login.

See also

As you can see, integrating a third-party authentication in the Authorization Server just requires configuring the client application for the IdP. So, if you need to integrate with another social provider, you will need to obtain the client application data.

If you want to integrate with GitHub, you can create an application registration on the `https://github.com/settings/developers` page.

For Facebook, you can create your application on the developer's page at `https://developers.facebook.com/apps`.

Integrating a RESTful API with a cloud IdP

Now that your application is becoming more and more popular, you decide to delegate your authentication to a cloud IdP as they offer advanced protection for sophisticated threats. You decide to use Azure AD B2C. This service is intended for public-facing applications, allowing customers to sign in and sign up, as well as customize the user journey, social network integration, and other interesting features.

What you'll learn in this recipe can be applied to other cloud IdPs, such as Okta, AWS Cognito, Google Firebase, and many others. Spring Boot offers specialized starters that simplify the process of integrating with an IdP even more.

Getting ready

In this recipe, we'll integrate the application you prepared in the *Configuring an MVC application with OpenID authentication* recipe. If you haven't completed that recipe yet, I've prepared a working version that you can find in this book's GitHub repository at `https://github.com/PacktPublishing/Spring-Boot-3.0-Cookbook`, in the `chapter4/recipe4-6/start` folder. This recipe also requires Redis, as explained in the *Configuring an MVC application with OpenID authentication* recipe. The easiest way to deploy it on your computer is by using Docker.

As we'll be integrating with Azure AD B2C, you will need an Azure subscription. If you don't have one, you can create a free account at `https://azure.microsoft.com/free`. Azure AD B2C offers a free tier that allows up to 50,000 monthly active users.

If you don't have an Azure AD B2C tenant, please follow the instructions at `https://learn.microsoft.com/azure/active-directory-b2c/tutorial-create-tenant` to create one before starting this recipe.

How to do it...

Follow these steps to build a smooth sign-in/signup process with Azure AD B2C and learn how to connect it seamlessly with your Spring Boot application for large-scale user authentication:

1. The first thing you'll need to do is create an application registration in Azure AD B2C. An application registration is the same as the client registration you created in Spring Authorization Server in previous recipes. You can create the application registration in the **App registrations** section:

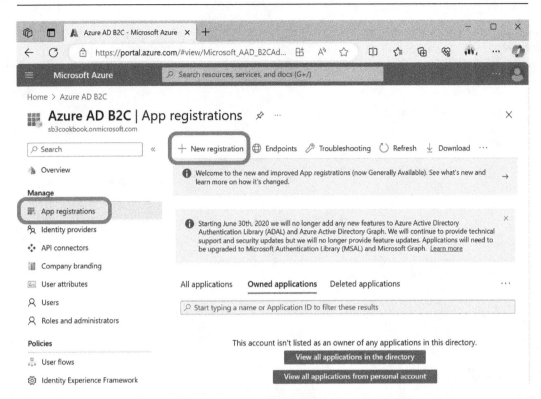

Figure 2.21: Creating an app registration in Azure AD B2C

2. On the **App registrations** page, set the following parameters:

    - For **Name**, set the name of the application. I suggest that you use `Football UI`.

    - For **Supported account types**, select **Accounts in any identity provider or organizational directory (for authenticating users with user flows)**. It's the default option.

    - For **Redirect URI (recommended)**, configure **Web** as the platform and `http://localhost:9080/login/oauth/code` as the value of the redirect UI:

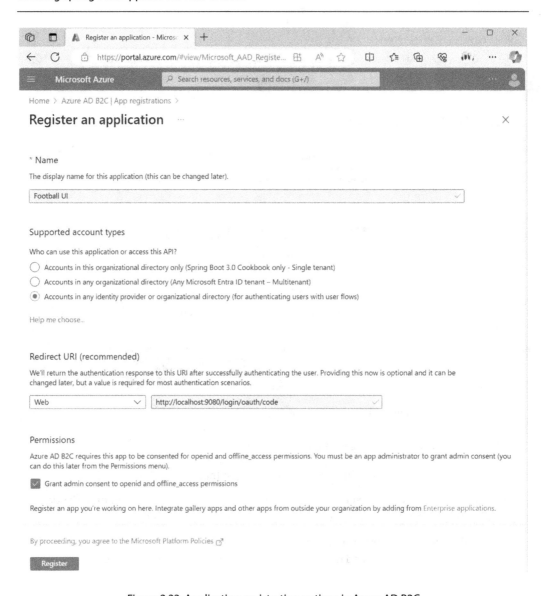

Figure 2.22: Application registration options in Azure AD B2C

Click **Register** to continue with the application registration process.

3. Once you have created the application registration, you need to configure a client secret. You can do this in the **Certificates & secrets** section of the application registration you've created:

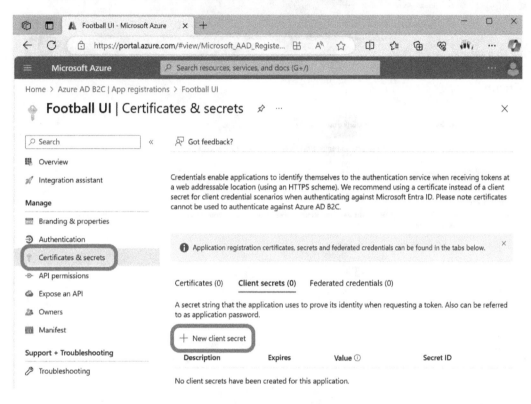

Figure 2.23: Creating a new client secret

4. Once you've created the secret, the secret value generated by Azure AD B2C will appear. You should copy the value now as it won't be accessible again. Keep it safe as you'll need it later.

5. Finally, we must create a user flow. A user flow is a configuration policy that can be used to set up the authentication experience for end users. Set the following options:

   - For **User flow type**, select **Sign up and sign in**. Set the version **Recommended**.

   - For **Name**, type SUSI; this is an acronym for "sign up and sign in."

   - For **Identity providers**, select **Email signup**.

   - For **User attributes and token claims**, select **Given Name** and **Surname**. For both attributes, check the **Collect attribute** and **Return claim** boxes.

- Keep the rest of the options as-is:

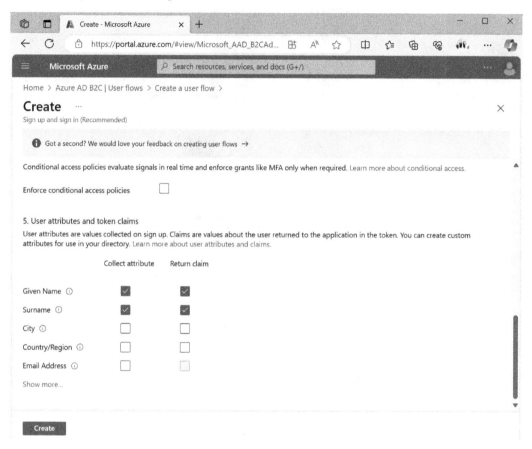

Figure 2.24: The Create a user flow page

Click **Create** to create the user flow.

6. Now, let's configure our application. First, you'll need to add the appropriate dependencies. For that, open the `pom.xml` file of the web application and add the `org.springframework.boot:spring-boot-starter-oauth2-client` dependency:

```
<dependency>
    <groupId>org.springframework.boot</groupId>
    <artifactId>spring-boot-starter-oauth2-client
        </artifactId>
</dependency>
```

7. Then, you'll need to configure the Azure AD B2C settings in the `application.yml` file in the `resources` folder. Replace the Oauth2 client settings with the B2C ones. The file should look as follows:

```
server:
  port: 9080
spring:
  cloud:
    azure:
      active-directory:
        b2c:
          enabled: true
          base-uri:
            https://sb3cookbook.b2clogin.com/
              sb3cookbook.onmicrosoft.com/
          credential:
            client-id:
              aa71b816-3d6e-4ee1-876b-83d5a60c4d84
            client-secret: '<the secret>'
          login-flow: sign-up-or-sign-in
          logout-success-url: http://localhost:9080
          user-flows:
            sign-up-or-sign-in: B2C_1_SUSI
          user-name-attribute-name: given_name
  data:
    redis:
      host: localhost
      port: 6379
```

In the `client-secret` field, set the value you kept in *Step 4*. I recommend enclosing the secret value in quotation marks since the secret likely contains reserved characters that can cause unexpected behavior when the `application.yaml` file is being processed.

8. To complete the OpenID configuration to allow users to log in to your application with Azure AD B2C, you'll need to adjust the security chain by applying an Azure AD OIDC configurer. To do this, modify the `SecurityConfiguration` class by adding `AadB2cOidcLoginConfigurer` in the constructor to allow bean injection, and then use it in the existing `defaultSecurityFilterChain` method, as follows:

```
private final AadB2cOidcLoginConfigurer configurer;
public SecurityConfiguration(AadB2cOidcLoginConfigurer
configurer) {
    this.configurer = configurer;
}
@Bean
```

```
public SecurityFilterChain
defaultSecurityFilterChain(HttpSecurity http) throws
Exception {
    http
        .authorizeHttpRequests((authorize) ->
            authorize
            .requestMatchers("/").permitAll()
            .anyRequest().authenticated())
        .apply(configurer);
    return http.build();
}
```

9. At this point, you can run your web application and authenticate with Azure AD B2C. However, there's still something pending that's protecting the RESTful API server with Azure AD B2C.

    To get around this, you can modify the dependencies. For that, open the pom.xml file of the RESTful API project and replace the org.springframework.boot: spring-boot-starter-oauth2-resource-server dependency with com.azure.spring:spring-cloud-azure-starter-active-directory-b2c:

    ```
    <dependency>
      <groupId>com.azure.spring</groupId>
      <artifactId>spring-cloud-azure-starter-active-
        directory-b2c</artifactId>
    </dependency>
    ```

10. Now, modify the application.yml file so that it configures the Azure B2C client registration:

    ```
    spring:
      cloud:
        azure:
          active-directory:
            b2c:
              enabled: true
              profile:
                tenant-id:
                  b2b8f451-385b-4b9d-9268-244a8f05b32f
              credential:
                client-id:
                  aa71b816-3d6e-4ee1-876b-83d5a60c4d84
              base-uri: https://sb3cookbook.b2clogin.com
              user-flows:
                sign-up-or-sign-in: B2C_1_SISU
    ```

11. Now, you can run both the web application and the RESTful server since both are protected with Azure AD B2C:

  I. Open your browser and navigate to `http://localhost:8080/myself`. As the method is protected, you will be redirected to the **Azure AD B2C Sign up or sign in** page:

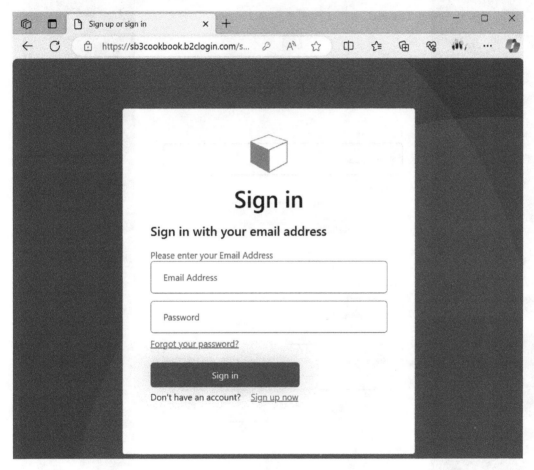

Figure 2.25: The Azure AD B2C default Sign up or sign in page

II.   If you click on the **Sign up now** link, you can create a new user:

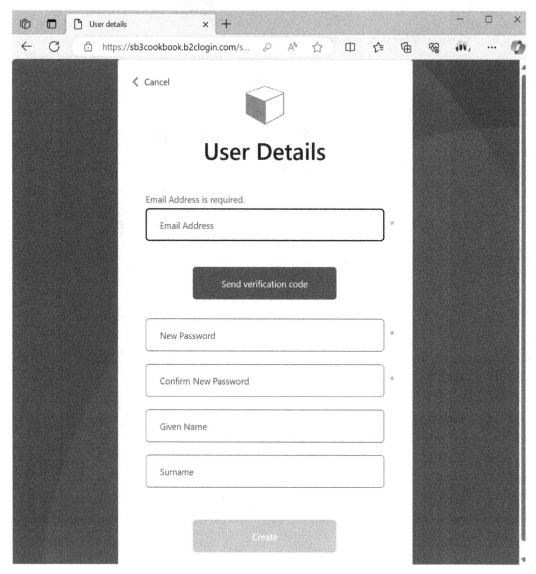

Figure 2.26: Sign-up page

III.   The first step is to provide a valid email address. Once you've done this click **Send verification code**. You will receive an email containing a verification code that you will need to provide on this page. Once verified, you can introduce the rest of the fields.

IV. When you return to the page, you will see the claims that have been provided by Azure AD B2C:

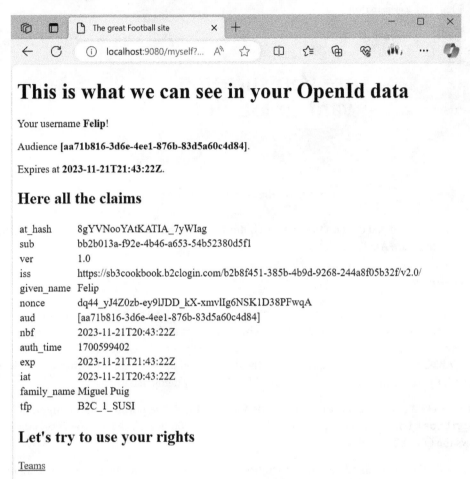

Figure 2.27: Our web page showing the claims from Azure B2C

V.   If you click on the **Teams** link, you will see the same data that you did in the *Configuring an MVC application with OpenID authentication* recipe.

VI.   If you click on the **Logout** link, you will be redirected to the default logout page:

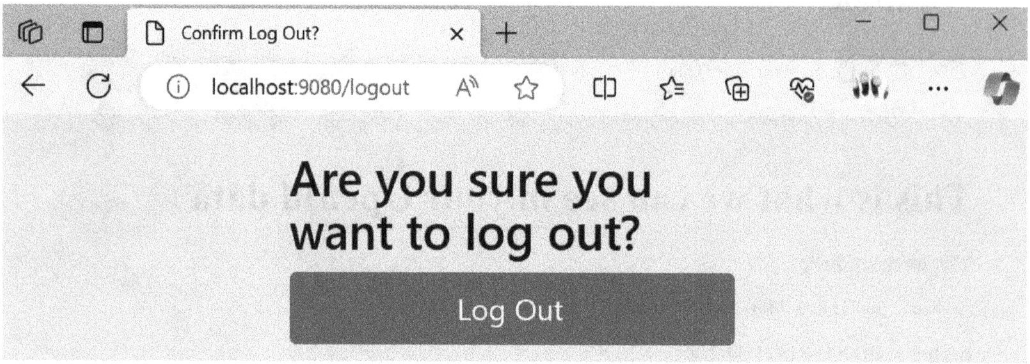

Figure 2.28: The default logout page

VII. Finally, if you click on the **Log Out** button, you will be redirected to the logout endpoint in Azure AD B2C.

How it works...

The `com.azure.spring:spring-cloud-azure-starter-active-directory-b2c` dependency includes both the Spring OAuth2 client and Spring OAuth2 resource starters that we used in previous recipes. On top of those starters, it also includes specific components to adapt the Azure AD B2C-specific features. For instance, the discovery endpoint cannot be inferred just from the issuer URL as it is dependent on the Azure AD B2C policy being used.

The Azure AD B2C starter maps the configuration used in the Azure portal to the configuration on your `application.yml` file. Apart from that, the application doesn't require specific changes as it follows the OAuth2 specification.

In Azure AD B2C, we defined an app registration. This is equivalent to the concept of a client in Spring Authorization Server.

Azure AD B2C allows us to define different policies so that we can customize the user experience. We created a policy that performs the sign-up and sign-in process using the default settings, but you can define an edit profile or reset password policy as well. Other interesting capabilities include defining custom interfaces and integrating other IdPs. For instance, it's quite easy to integrate with cloud providers such as Google Accounts, social network providers such as Facebook and Instagram, and enterprise IdPs such as Azure Entra.

One of the main advantages of this solution is that users can register themselves; they don't need an administrator to do this for them.

This recipe doesn't intend to review all the possibilities of Azure AD B2C – it's been provided to help you understand how to integrate your Spring Boot application with Azure AD B2C.

There's more...

An interesting and probably more frequent scenario is when our RESTful API uses a different application registration than the UI application. When I build a RESTful API, I usually design it while keeping one thing in mind: more than one client should be able to consume it. This works for different scenarios, such as the web and mobile versions, or allowing a third-party application to consume some of the APIs. In that scenario, you can create a dedicated application registration for your RESTful API and create different application roles with different access levels. Then, you can assign the corresponding role to the consumer applications.

When you create the application registration for the RESTful API, you can create the roles by opening the manifest and including the desired roles of your application. For instance, we can create the `football.read` role for general consumer access and the `football.admin` role for administrative access. It would look like this:

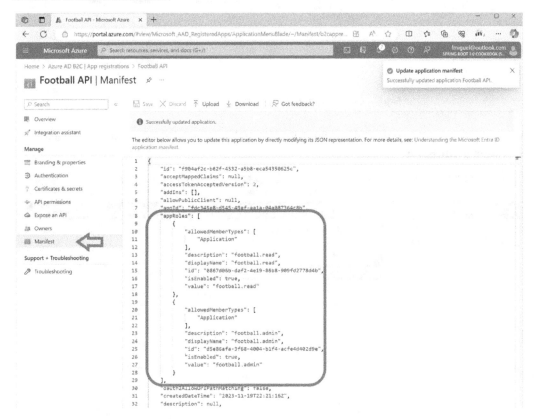

Figure 2.29: The application registration manifest with two application roles

Then, in the RESTful application registration area, go to **Expose an API** and assign an **Application ID URI** value:

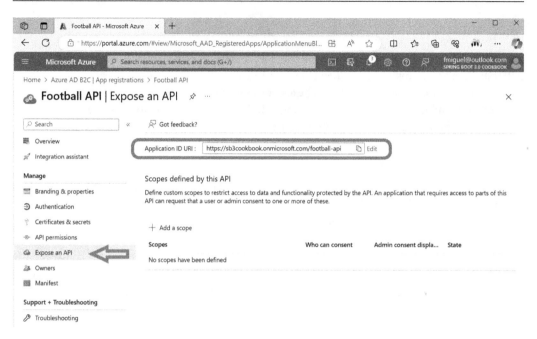

Figure 2.30: Assigning an Application ID URI value

Then, we can assign permissions to the RESTful API. Go to **API permissions** and assign **Application permissions**:

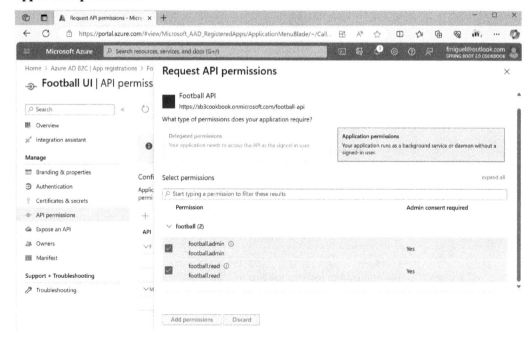

Figure 2.31: Assigning application permissions to the consumer application

Now, the RESTful API has its own application registration. This means that you can configure the application with its own audience, instead of the UI application. To configure that, go to the `application.yml` file of the RESTful API and change the `client-id` property to the client ID of the application registration of the RESTful API.

If you want to use the application roles to provide different access levels, you will need to use `AadJwtGrantedAuthoritiesConverter` from the Azure AD B2C starter. You can register the bean in the `SecurityConfig` class, as follows:

```
@Bean
public Converter<Jwt, Collection<GrantedAuthority>>
aadJwtGrantedAuthoritiesConverter() {
    return new AadJwtGrantedAuthoritiesConverter();
}
@Bean
public JwtAuthenticationConverter
aadJwtAuthenticationConverter() {
    JwtAuthenticationConverter converter = new
        JwtAuthenticationConverter();
        converter.setJwtGrantedAuthoritiesConverter(
            aadJwtGrantedAuthoritiesConverter());
    return converter;
}
```

By default, the Spring OAuth2 resource server only converts the `scope` claim and the application roles will be provided in the `roles` claim. The converter generates authorities for each role with the `APPROLE_` prefix. So, you can restrict access using these authorities like so:

```
@Bean
public SecurityFilterChain filterChain(HttpSecurity http) throws
Exception {
    return http
        .authorizeHttpRequests(authorize -> authorize
            .requestMatchers(HttpMethod.GET,
                "/football/teams/**").hasAnyAuthority(
                "APPROLE_football.read",
                "APPROLE_football.admin")
            .requestMatchers(HttpMethod.POST,
                "/football/teams/**").hasAnyAuthority(
                "APPROLE_football.admin")
            .anyRequest().authenticated())
        .oauth2ResourceServer(oauth2 ->
            oauth2.jwt(Customizer.withDefaults()))
        .build();
}
```

Here's what the payload of an access token with the application roles looks like:

```
{
  «aud»: «fdc345e8-d545-49af-aa1a-04a087364c8b»,
  "iss": "https://login.microsoftonline.com/b2b8f451-385b-
    4b9d-9268-244a8f05b32f/v2.0",
  "iat": 1700518483,
  "nbf": 1700518483,
  "exp": 1700522383,
  "aio": "ASQA2/8VAAAAIXIjK+
    28DPOc4epV22pKGfqdRSnps2dtReyZY7MPhpk=",
  "azp": "aa71b816-3d6e-4ee1-876b-83d5a60c4d84",
  "azpacr": "1",
  "oid": "d88d83d6-421f-41e2-ba99-f49516fd439a",
  "rh": "0.ASQAUfS4sls4nUuSaCRKjwWzL-hFw_
    1F1a9JqhoEoIc2TIskAAA.",
  «roles»: [
    "football.read",
    "football.admin"
  ],
  "sub": "d88d83d6-421f-41e2-ba99-f49516fd439a",
  "tid": "b2b8f451-385b-4b9d-9268-244a8f05b32f",
  "uti": "JSxYHbHkpUS91mwBtxNaAA",
  "ver": "2.0"
}
```

By doing this, only the allowed client applications will be able to consume the RESTful API.

Observability, Monitoring, and Application Management

Monitoring and observability are crucial aspects of managing and maintaining the health, performance, and reliability of modern applications. In microservices-oriented applications, with multiple instances of different services running at the same time to deliver a solution, observability and monitoring help in understanding the interactions between these services and identifying issues.

Monitoring plays a crucial role in large environments, enabling resource utilization and performance metrics to be tracked. This, in turn, facilitates dynamically scaling resources to effectively meet the demands of the system. This is especially useful in cloud computing environments, where you pay for the resources used and where you can adapt your application resources to the real demands of your users. Without monitoring, how do you know if your application is running at 100% CPU and the response time is so slow that your users abandon your application?

When you have multiple microservices running in your application and there's an issue, observability is crucial in identifying the failing component and the context in which errors occur.

Observability and monitoring are also very important for continuous improvement. You can use the insights gained from monitoring to make data-driven decisions, enhance performance, and refine the solution over time.

Spring Boot, through Actuator, provides not only monitoring but also management capabilities that allow you to interact with the application in production environments. This capability not only allows you to detect potential issues in the application but also helps in troubleshooting at runtime.

In this chapter, you will gain insights into activating observability and monitoring features within your Spring Boot applications. We'll start by providing health checks in your application. Here, you'll learn how to leverage the data that's generated by your application through popular open source solutions. This chapter will also cover creating traces within your system, allowing you to correlate activities across different microservices and explore them using Zipkin. Additionally, you will learn how to monitor the exposed metrics of your application using Prometheus and Grafana. Beyond the standard metrics provided by Spring Boot and its associated components, you will generate custom metrics tailored to your application's specifics and monitor them. Once your application becomes both monitorable and observable, you can also integrate with commercial tools while considering the plethora of powerful monitoring solutions available in the market that are well-suited for production environments. Finally, you will learn how to change application settings at runtime so that you can troubleshoot your application.

In this chapter, we're going to cover the following recipes:

- Adding Actuator to your application
- Creating a custom Actuator endpoint
- Using probes and creating a custom health check
- Implementing distributed tracing
- Accessing standard metrics
- Creating your own metrics
- Integrating your application with Prometheus and Grafana
- Changing the settings of a running application

Technical requirements

In this chapter, we'll need to run different tools, such as Prometheus, Grafana, and Zipkin. As usual, the simplest way to run them on your computer is using Docker. You can get Docker from its product page: `https://www.docker.com/products/docker-desktop/`. I will explain how to deploy each tool in its corresponding recipe.

All the recipes that will be demonstrated in this chapter can be found at: `https://github.com/PacktPublishing/Spring-Boot-3.0-Cookbook/tree/main/chapter3`.

Adding Actuator to your application

So, you plan to develop a new RESTful API to complete your football-related suite of services. You are concerned about the responsiveness of your application and our aim to provide a resilient service. For that reason, you are very interested in monitoring your application health properly.

Before you start to monitor your application, your application should be monitorable. For that, you have decided to start using **Spring Boot Actuator**.

Spring Boot Actuator comprises a set of production-ready functionalities packaged with the Spring Framework. It incorporates various built-in tools and endpoints that are designed to allow you to monitor, manage, and interact with Spring Boot applications within a production setting. Actuator simplifies the process of comprehending and resolving runtime behaviors in Spring Boot applications.

The Actuator module exposes multiple endpoints, including `health`, `metrics`, `info`, `dump`, and `env`, among others, offering operational insights into the running application. Once this dependency is included, you have a lot of out-of-the-box endpoints available. Customizing and extending these endpoints can easily be achieved and provides flexibility in terms of configuration.

In this recipe, you will learn how to include Spring Boot Actuator in your project and use some of the endpoints that are provided out of the box.

Getting ready

In this recipe, we'll create an application using the *Spring Initializr* tool. As you did in previous chapters of this book, you can use the tool in your browser by going to `https://start.spring.io` or integrating it into your favorite code editor.

How to do it...

Let's create a project with Actuator enabled and start exploring the endpoints provided:

1. Create a project using the *Spring Initializr* tool. Open `https://start.spring.io` and use the same parameters that you used in the *Creating a RESTful API recipe* of *Chapter 1*, except change the following options:

    *   For **Artifact**, type `fooballobs`

    *   For **Dependencies**, select **Spring Web** and **Spring Boot Actuator**

2. Download the template that was generated with *Spring Initializr* and unzip the content to your working directory.

3. If you run the application now, you can access the health endpoint at `/actuator/health`. Before running the application, we'll expose some endpoints. For that, create an `application.yml` file in the `resources` folder and add the following content:

    ```
    management:
        endpoints:
            web:
                exposure:
                    include: health,env,metrics,beans,loggers
    ```

4. Now, run the application. You can now access the exposed Actuator endpoints:

    *   `http://localhost:8080/actuator/health`: This endpoint provides health information about your application. It is very useful in containerized environments such as Kubernetes to ensure that your application is up and running.

    *   `http://localhost:8080/actuator/env`: This endpoint returns the environment variables of the application.

- `http://localhost:8080/actuator/metrics`: This endpoint returns a list that contains the metrics that have been exposed by the application. You can get the values of any of the metrics that have been exposed by appending the name to the metrics endpoint. For instance, to get `process.cpu.usage`, you can request `http://localhost:8080/actuator/metrics/process.cpu.usage`.

- `http://localhost:8080/actuator/beans`: This endpoint returns a list with the beans registered in the IoC container – that is, a list of beans that can be injected into other beans.

- `http://localhost:8080/actuator/loggers`: This endpoint returns the list of log levels and loggers of the application. It also allows to modify the log level at runtime.

5. In this recipe, you exposed just some of the available endpoints. You can find the full list of built-in endpoints at `https://docs.spring.io/spring-boot/docs/current/reference/html/actuator.html#actuator.endpoints`.

How it works...

When you integrate Actuator into your application, it provides a set of endpoints that can be used for monitoring your application and managing its behavior. In addition to the built-in endpoints, it lets you add your own.

Endpoints can be enabled or disabled. By default, all endpoints are enabled except the shutdown endpoint – as its name suggests, you can use it to gracefully shut down the application. Then, the endpoints can be exposed, meaning that they can be accessed remotely using HTTP requests or JMX. By default, only the health endpoint is exposed. In this book, we'll mostly focus on HTTP as it can be used with standard monitoring tools not specific to the Java ecosystem. HTTP is only available for web applications; if you're developing another type of application, you will need to use JMX.

Depending on the components you use, more data will be exposed. For instance, when you include Spring Data JPA, the Spring Data metrics become available, so you will have to configure the number of open connections and other relevant metrics for Spring Data monitoring.

There's more...

Some of the endpoints provided by Actuator may expose very sensitive information. So, the health endpoint is the only one that's exposed by default. If your applications can only be accessed inside a virtual network or protected with a firewall, maybe you can keep endpoints open. Whether your application is publicly exposed or you simply want to control who accesses your Actuator endpoint, you may want to protect them, as explained in *Chapter 2*. For instance, a security configuration could look as follows:

```
@Configuration
public class SecurityConfig {
```

```
    @Bean
    public SecurityFilterChain filterChain(HttpSecurity http) throws
Exception {
        return http
                .authorizeHttpRequests(authorize ->
                    authorize.requestMatchers("/actuator/**")
                    .hasRole("ADMIN")
                    .anyRequest().authenticated())
                .oauth2ResourceServer(oauth2 -> oauth2.jwt(Customizer.
withDefaults()))
                .build();
    }
}
```

You can refer to Spring Boot's official documentation at https://docs.spring.io/spring-boot/docs/current/reference/html/actuator.html#actuator.endpoints.security for more details.

See also

In addition to endpoints provided by Spring Boot and the components used, Actuator provides a flexible implementation that allows you to create your own endpoints. Later in this chapter, you will learn how to create your own Actuator endpoint, metrics, and custom health checks.

See the *Creating a custom Actuator endpoint*, *Creating a custom health check*, and *Creating your own metrics* recipes for more information.

Creating a custom Actuator endpoint

In our example, we are developing a new RESTful API that requires a file to be loaded from blob storage. That file doesn't change frequently, which means it's loaded in memory at application startup and is not reloaded again automatically. You need to know which version of the file is loaded, and you want to force a reload when there is a new version.

To implement this feature, you will use a custom Actuator endpoint. This endpoint will have a GET operation to return the current file version, and a POST method to reload the file.

Getting ready

In this recipe, you will reuse the application you created in the *Adding Actuator to your application* recipe. I've prepared a working version in this book's GitHub repository at https://github.com/PacktPublishing/Spring-Boot-3.0-Cookbook/. It can be found in the chapter3/recipe3-2/start folder.

How to do it...

Let's modify the RESTful API so that it loads a file from a folder and returns some results. Once you've done this, you'll need to create a custom Actuator endpoint that returns the file that's been loaded. You will also need to configure the endpoint to reload the file:

1. Start by creating a class that loads a file and keeps the content in memory:

   I.   Let's name it `FileLoader` and add the following code:

   ```
   public class FileLoader {
       private String fileName;
       private List<String> teams;
       private String folder;
       public FileLoader(String folder) {
           this.folder = folder;
       }
   }
   ```

   II. To load the file and keep the content in memory, add the following code:

   ```
   private void loadFile(String fileName) throws Exception {
           this.fileName = fileName;
           ObjectMapper mapper = new ObjectMapper();
           File file = new File(fileName);
           teams = mapper.readValue(file,
               new TypeReference<List<String>>() {
               });
   }
   ```

   III. Now, add a public method so that you can load the first file that's found in the folder that's passed in the constructor:

   ```
   public void loadFile() throws IOException {
       Files.list(Paths.get(folder))
           .filter(Files::isRegularFile)
           .findFirst()
           .ifPresent(file -> {
               try {
                   loadFile(file.toString());
               } catch (Exception e) {
                   e.printStackTrace();
               }
           });
   }
   ```

2. Next, create a class annotated with @Endpoint to define the custom Actuator endpoint. Name it FootballCustomEndpoint:

```
@Endpoint(id = "football")
public class FootballCustomEndpoint {
    private FileLoader fileLoader;
    FootballCustomEndpoint(FileLoader fileLoader){
        this.fileLoader = fileLoader;
    }
}
```

This class receives a FileLoader object in the constructor to perform the necessary actions.

3. Now, create the custom endpoint operations in FootballCustomEndpoint:

I. Create a method annotated with @ReadOperation to retrieve the file version in use:

```
@ReadOperation
public String getFileVersion(){
    return fileLoader.getFileName();
}
```

II. Create a method annotated with @WriteOperation to refresh the file:

```
@WriteOperation
public void refreshFile(){
    try {
        fileLoader.loadFile();
    } catch (Exception e) {
        e.printStackTrace();
    }
}
```

4. Next, you need to create a bean for both the FileLoader and FootballCustom Endpoint classes:

I. Create a class named FootballConfiguration and annotate it with @ Configuration:

```
@Configuration
public class FootballConfiguration {
    @Value("${football.folder}")
    private String folder;
}
```

II. Note that there is a field annotated with @Value. It will load the folder path containing the file to load from the configuration.

III.   Create a method that produces a bean for `FileLoader`:

```
@Bean
public FileLoader fileLoader() throws IOException{
    FileLoader fileLoader = new FileLoader(folder);
    return fileLoader;
}
```

IV.   Now, create a method that produces `FootballCustomEndpoint`:

```
@Bean
public FootballCustomEndpoint footballCustomEndpoint(FileLoader
fileLoader){
    return new FootballCustomEndpoint(fileLoader);
}
```

5.   Since `FileLoader` needs to load the file by using the `loadFile` method, you will need to create a class that implement an `ApplicationRunner` interface:

```
@Component
public class DataInitializer implements ApplicationRunner {
    private FileLoader fileLoader;
    public DataInitializer(FileLoader fileLoader) {
        this.fileLoader = fileLoader;
    }
    @Override
    public void run(ApplicationArguments args) throws Exception
{
        fileLoader.loadFile();
    }
}
```

6.   Modify the `application.yml` file in the `resources` folder:

I.   Add a setting that provides a path to the folder containing the file to load:

```
football:
    folder: teams
```

II.   Add the new Actuator endpoint:

```
management:
    endpoints:
        web:
            exposure:
                include:
health,env,metrics,beans,loggers,football
```

7. Finally, create a folder named `teams` in the root of the project and a file named `1.0.0.json`. As content, add an array containing teams – for example, `[ "Argentina", "Australia", "Brazil"]`.

8. Create a sample RESTful controller that returns the content that's loaded in memory by the `FileLoader` class:

```
@RestController
@RequestMapping("/football")
public class FootballController {
    private FileLoader fileLoader;
    public FootballController(FileLoader fileLoader){
        this.fileLoader = fileLoader;
    }
    @GetMapping
    public List<String> getTeams(){
        return fileLoader.getTeams();
    }
}
```

9. The service is now ready to test. Execute the application and perform the following requests:

    I.   Get the current file version using the custom Actuator endpoint. For that, open your terminal and execute the following `curl` request:

    ```
    curl http://localhost:8080/actuator/football
    ```

    II. You will receive the filename as a response – that is, `teams/1.0.0.json`.

    III. Let's create a new version of the file. Rename the file `1.0.1.json` and add a new element to the `teams` array, like so:

    ```
    [ "Senegal", "Argentina", "Australia", "Brazil"]
    ```

    IV. Now, use the custom Actuator endpoint to refresh the file in the application. For that, in your terminal, execute the following `curl` request:

    ```
    curl --request POST http://localhost:8080/actuator/football
    ```

    V.   Check the current file version again; you will now get `teams/1.0.1.json`.

    VI. You can also use a RESTful API to validate that the results correspond with the content of the file.

How it works...

By creating a bean with the `@Endpoint` annotation, Actuator exposes all methods annotated with `@ReadOperation`, `@WriteOperation`, and `@DeleteOperation` over JMX and HTTP. This example is not much different from a regular RESTful endpoint, but the purpose is different as it's used to manage the application or library you developed. Of course, you can implement your custom Actuator endpoint, but usually, Actuator endpoints are provided as part of a component that is used by others and may require some internal information or behavior to be exposed. For instance, database drivers such as PostgreSQL, database connection pool managers such as HikariCP, and caching systems such as Redis usually provide Actuator endpoints. If you plan to create some kind of system or library that will be used by others and you are interested in exposing some internals to facilitate management in runtime, Actuator endpoints are a great solution.

An `ApplicationRunner` is a component that is executed right after the application starts. When Spring Boot executes, the `ApplicationRunner` isn't ready to accept requests yet. You can define more than one `ApplicationRunner`. Once all the `ApplicationRunner` components are executed, the application is ready to accept requests.

Using probes and creating a custom health check

Your new football trading service is getting readily adopted by football fans. This service is used to exchange stickers with football players' pictures on them between fans. To accelerate the process, the service caches some data in the application's memory. You need to ensure that the cache is filled before you start serving requests.

Under normal conditions, the football trading service works fine; however, under heavy load, the application instances start degrading and after some instability, they end up being unresponsive. To counteract this, you prepare some stress tests in the lab environment. However, you realize that the application starts degrading because you have issues connecting to the database. At the same time, you realize that those kinds of issues happen when the application has more than 90 pending orders. While you find a definitive solution, you decide to expose when the application is unable to process more requests and create a health check that verifies if it can connect to the database.

Probes are mostly used by container orchestrators, such as Kubernetes, to verify that the application is ready to accept requests and when it is already working to indicate that it's alive. In Kubernetes, they are known as readiness and liveness probes.

A health check is a mechanism to verify that the application has everything ready to work – for instance, it's able to connect to a database.

In this recipe, you will learn how to expose a readiness check, how to change your liveness state, and how to create a custom health check that can be used by the hosting platform or a monitoring system to determine the health of your application instances and when your application is ready to accept requests.

Getting ready

In this recipe, you will reuse the application you created in the *Creating a custom Actuator endpoint* recipe. I've prepared a working version in this book's GitHub repository at `https://github.com/PacktPublishing/Spring-Boot-3.0-Cookbook/`. It can be found in the `chapter3/recipe3-3/start` folder.

In this recipe, you will verify that the application can connect to the application database. We'll use PostgreSQL as a database. To run PostgreSQL locally, we'll use Docker. You can download and start the database just by executing the following command in your terminal:

```
docker run -e POSTGRES_USER=packt -e POSTGRES_PASSWORD=packt -p
5432:5432 --name postgresql postgres
```

If you created any database in *Chapter 5*, you can reuse it here. This recipe doesn't perform any real queries – it just verifies it can connect. If you don't have a database created in the container, you can create a database using the **psql** tool. For that, execute the following command in your terminal:

```
psql -h localhost -U packt
```

You will be prompted for a password. Specify `packt` and press **intro**. You will be connected to a PostgreSQL terminal. Execute the following command to create a database:

```
CREATE DATABASE football;
```

Now, you can exit the database by executing the `quit;` command.

How to do it...

In this recipe, you'll configure your application so that it can manage probes and create a custom health check to verify that the application can connect to the database:

1. Start by updating the `application.yml` file in the `resources` folder so that it can enable readiness and liveness probes. For that, include the following:

    ```yaml
    management:
        endpoint:
            health:
                probes:
                    enabled: true
    ```

2. Next, create a class that emulates the football trading service. Name it `TradingService`:

    ```java
    @Service
    public class TradingService {
    }
    ```

This class will manage the trading requests. When trading a request, if it detects that there are more than 90 pending orders, it will notify you that the application cannot manage more requests. For that, it will use `ApplicationEventPublisher`, which will be injected into the constructor:

```
private ApplicationEventPublisher applicationEventPublisher;
public TradingService(ApplicationEventPublisher
applicationEventPublisher) {
    this.applicationEventPublisher = applicationEventPublisher;
}
```

Next, define a method that returns the number of pending orders. We'll simulate this by returning a random number between 0 and 100:

```
public int getPendingOrders() {
    Random random = new Random();
    return random.nextInt(100);
}
```

Finally, you can create a method that manages the trading operations. If there are more than 90 pending orders, it will change the state of the application:

```
public int tradeCards(int orders) {
    if (getPendingOrders() > 90) {
        AvailabilityChangeEvent.
publish(applicationEventPublisher, new Exception("There are more
than 90 pending orders"), LivenessState.BROKEN);
    } else {
        AvailabilityChangeEvent.
publish(applicationEventPublisher, new Exception("working
fine"), LivenessState.CORRECT);
    }
    return orders;
}
```

3. Now, configure the connection to the database:

I. Add Spring Data JDBC and PostgreSQL dependencies. For that, in the `pom.xml` file, add the following dependencies:

```
<dependency>
    <groupId>org.springframework.boot</groupId>
    <artifactId>spring-boot-starter-data-jdbc</artifactId>
</dependency>
<dependency>
    <groupId>org.postgresql</groupId>
    <artifactId>postgresql</artifactId>
    <scope>runtime</scope>
</dependency>
```

II. Add the following configuration to the `application.yml` file in the `resources` folder:

```
spring:
    datasource:
        url: jdbc:postgresql://localhost:5432/football
        username: packt
        password: packt
```

4. Now, let's create a health indicator:

I. For that, create a class named `FootballHealthIndicator` that implements the `HealthIndicator` interface:

```
@Component
public class FootballHealthIndicator implements HealthIndicator
{
}
```

II. As it will connect to the database, inject `JdbcTemplate` into the constructor:

```
private JdbcTemplate template;
public FootballHealthIndicator(JdbcTemplate template) {
    this.template = template;
}
```

III. Now, override the health method so that you can perform connectivity checking:

```
@Override
public Health health() {
    try {
        template.execute("SELECT 1");
        return Health.up().build();
    } catch (DataAccessException e) {
        return Health.down().withDetail("Cannot connect to
database", e).build();
    }
}
```

5. Before testing the application, you can modify the `FileLoader` class, simulating it so that it takes a few seconds to load the file. You can do this by modifying the `loadFile` method by adding the following code. This will make the application wait 10 seconds before it loads the file:

```
try {
    Thread.sleep(10000);
} catch (InterruptedException e) {
    e.printStackTrace();
}
```

6. Now, let's test the application's readiness:

I.   Before running the application, execute the following command in your terminal:

```
watch curl http://localhost:8080/actuator/health/readiness
```

This command will execute a request to the readiness probe every second.

II.   Start the application. You will see that the output of the watch command changes. First, it will appear as **OUT_OF_SERVICE**:

Figure 3.1: Readiness status set to OUT_OF_SERVICE

III. After 10 seconds or the time you configured in *Step 5*, it will change to **UP**:

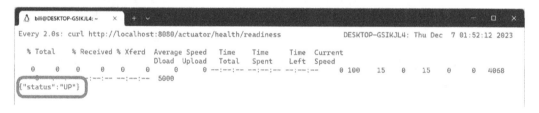

Figure 3.2: Readiness status changed to UP

7. Now, test the application's liveness:

I.   Again, execute a watch command, but this time, make requests to the liveness probe's endpoint:

```
watch curl http://localhost:8080/actuator/health/readiness
```

You will see that the status is **UP**.

II.   Let's start using the trading endpoint. Run another watcher that executes trading requests. For that, execute the following command in a new terminal:

```
watch -n 1 -x curl --request POST -H "Content-Type: application/
json" --data "1" http://localhost:8080/football
```

Remember that if there are more than 90 pending requests, it will mark itself as failing. Since a random number between 0 and 100 is selected, there's a 10% possibility it will fail.

You will see that the readiness endpoint returns **DOWN** from time to time.

8.   Finally, test the health check endpoint:

I.   Execute a `watch` command for the Actuator health endpoint:

```
watch curl http://localhost:8080/actuator/health
```

II.   it will return **UP** every time. To verify that it detects when it cannot connect to the database, stop the PostgreSQL container. To do so, run the following command:

```
docker stop postgresql
```

You will see that the Actuator endpoint will take longer to respond and that the response will be **DOWN**.

How it works...

Readiness and liveness probes are enabled automatically when Spring Boot detects it's running on Kubernetes, but you can enable them manually. In this recipe, we enabled them explicitly, but if you run the application on Kubernetes, this will be done automatically.

Readiness and liveness probes should not check any external component. They should verify that the application is ready internally and that it's capable of responding. On the other hand, health checks should verify that all dependent components are available.

The Spring Boot application life cycle goes through different states, and it generates events every time it changes its state. I won't explain all possible application states here; instead, I'll focus on the relevant states during readiness probes. The first state is `starting`. Once Spring Boot initializes the components, it changes to `started`. At this point, it's not ready yet, so it needs to run all `ApplicationRunner` and `CommandLineRunner` instances defined in the application. Once all of them are executed, it changes to `ready`. In this recipe, we introduced a delay of 10 seconds in the `loadFile` method. During this period, the readiness status was **OUT_OF_SERVICE**. Once it had loaded the file, it changed to **UP**.

If you want to learn more, take a look at the following Spring Boot documentation: `https://docs.spring.io/spring-boot/docs/current/reference/html/features.html#features.spring-application.application-events-and-listeners`.

Be careful while checking other components. First, if it is another service, such as the one we created in this recipe, it will likely also have probes and health checks. Checking this via your service can be redundant. Second, try to make light checks; otherwise, you may generate too much load, which can cause performance issues. In this recipe, the SQL command we used was `SELECT 1`. This command connects to the database but doesn't require computing resources from the database engine out of the connection itself.

See also

Health checks should not necessarily imply that you check the health of all the dependencies of your application. Rather, you should check if your application has any problems that could be solved by reducing the load or by rebooting. If your application depends on an unresponsive service and you mark your application as unhealthy, the application instance will be restarted. However, if your problem is in another application, the problem won't disappear, and the application will be restarted again and again without solving the problem. For that kind of scenario, consider implementing a *circuit breaker* solution. See `https://spring.io/guides/gs/cloud-circuit-breaker/` for guidance on how to implement this using Spring Cloud.

Implementing distributed tracing

So far, you've created a solution with two microservices, the football trading microservice and the client microservice. Among other features, the trading microservice provides the ranking of players. The client microservice enhances the list of players by adding the ranking that was obtained from the trading microservice.

Distributed tracing emerges as a crucial tool as it offers a systematic approach to monitoring, analyzing, and optimizing the flow of requests between microservices. Distributed tracing is a method of monitoring and visualizing the flow of requests as they propagate through various components of a distributed system, providing insights into performance, latency, and dependencies between services.

In this recipe, you will learn how to enable distributed tracing for your microservices, export the data to Zipkin, and access the results.

Zipkin is an open source distributed tracing system that helps developers trace, monitor, and visualize the paths of requests as they travel through various microservices in a distributed system, providing valuable insights into performance and dependencies. What you will learn about Zipkin in this recipe can be easily adapted to other tools.

Getting ready

In this recipe, we'll visualize the traces using Zipkin. You can deploy it on your computer using Docker. For that, open your terminal and execute the following command:

```
docker run -d -p 9411:9411 openzipkin/zipkin
```

The preceding command will download an image with an OpenZipkin server, if you don't have one already, and start the server.

We'll reuse the trading service we created in the *Using probes and creating a custom health check* recipe. If you haven't completed it yet, don't worry – I've prepared a working version in this book's GitHub repository at `https://github.com/PacktPublishing/Spring-Boot-3.0-Cookbook/`. It can be found in the `chapter3/recipe3-4/start` folder.

How to do it...

Let's enable distributed tracing in the existing trading service and create the new client service. For the new client service, we'll need to ensure that distributed tracing is enabled as well. Before starting, ensure that your OpenZipkin server is running, as explained in the *Getting ready* section:

1. Start by enabling distributed tracing in the trading microservice you created in the *Using probes and creating a custom health check* recipe:

    I.   For that, open the pom.xml file and add the following dependencies:

    ```
    <dependency>
        <groupId>io.micrometer</groupId>
        <artifactId>micrometer-tracing-bridge-otel</artifactId>
    </dependency>
    <dependency>
        <groupId>io.opentelemetry</groupId>
        <artifactId>opentelemetry-exporter-zipkin</artifactId>
    </dependency>
    ```

    II.   The first dependency is a bridge between **Micrometer** and **OpenTelemetry**. The second dependency is an exporter from OpenTelemetry to **Zipkin**. I'll explain this in more detail in the *How it works...* section.

    III. Now the application can send the traces to Zipkin. However, before running the application, you'll need to make some adjustments. Open the application.yml file in the resources folder and add the following setting:

    ```
    management
        tracing:
            sampling:
                probability: 1.0
    ```

    IV.   By default, sampling is only set to 10%. This means that only 10% of traces are sent. With this change, you will send 100% of the traces.

    V.   In the same application.yml file, add the following configuration:

    ```
    spring:
        application:
            name: trading-service
    ```

This change is not mandatory but helps identify the service in distributed tracing.

2. Next, create the ranking endpoint in the football trading microservice that will be consumed by the client microservice. For that, in `FootballController`, create the following method:

```
@GetMapping("ranking/{player}")
public int getRanking(@PathVariable String player) {
    logger.info(«Preparing ranking for player {}», player);
    if (random.nextInt(100) > 97) {
        throw new RuntimeException("It's not possible to get the
ranking for player " + player
            + " at this moment. Please try again later.");
    }
    return random.nextInt(1000);
}
```

To simulate random errors, this method throws an exception when a random number from 0 to 99 is greater than 97 – that is, 2% of the time.

3. Next, create a new application that will act as the client application. As usual, you can create the template using the *Spring Initializr* tool:

    *   Open https://start.spring.io and use the same parameters that you did in the *Creating a RESTful API* recipe of *Chapter 1*, except change the following options:

        *   For **Artifact**, type `fooballclient`

        *   For **Dependencies**, select **Spring Web** and **Spring Boot Actuator**

    *   Add dependencies for OpenTelemetry and Zipkin, as you did for the football trading service application. So, open the pom.xml file and add the following dependencies:

```
<dependency>
    <groupId>io.micrometer</groupId>
    <artifactId>micrometer-tracing-bridge-otel</artifactId>
</dependency>
<dependency>
    <groupId>io.opentelemetry</groupId>
    <artifactId>opentelemetry-exporter-zipkin</artifactId>
</dependency>
```

4. In the client application, add a RESTful controller:

    I.   Name it `PlayersController`:

```
@RestController
@RequestMapping("/players")
public class PlayersController {
}
```

II.   This application must call the trading service. For that, it will use RestTemplate. To achieve the correlation between service calls, you should use RestTemplateBuilder to create RestTemplate. Then, inject RestTemplateBuilder into the controller's constructor:

```
private RestTemplate restTemplate;
public PlayersController (RestTemplateBuilder
restTemplateBuilder) {
    this.restTemplate = restTemplateBuilder.build();
}
```

III.   Now, you can create the controller method that calls the trading service of the other application:

```
@GetMapping
public List<PlayerRanking> getPlayers() {
    String url = "http://localhost:8080/football/ranking";
    List<String> players = List.of("Aitana Bonmatí", "Alexia
Putellas", "Andrea Falcón");
    return players.stream().map(player -> {
        int ranking = this.restTemplate.getForObject(url + "/" +
player, int.class);
        return new PlayerRanking(player, ranking);
    }).collect(Collectors.toList());
}
```

5.   Configure client application tracing in the application.yml file:

```
management:
    tracing:
        sampling:
            probability: 1.0
spring:
    application:
        name: football-client
```

As you did in the trading service, you should set sampling to 1.0 so that 100% of the traces will be recorded. To distinguish the client application from the trading service application, set the spring.application.name property to football-client.

6.   To avoid port conflicts with the trading application, configure the client application so that it uses port 8090. To do that, add the following parameter to the application.yml file:

```
server:
    port: 8090
```

7. Now, you can test the application. Call the client application; it will make multiple calls to the trading service. To make continuous requests to the client application, you can execute the following command in your terminal:

```
watch curl http://localhost:8090/players
```

8. Finally, open Zipkin to see the traces. For that, go to `http://localhost:9411/` in your browser:

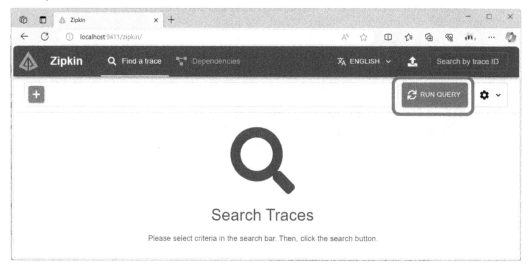

Figure 3.3: The Zipkin home page

On the home page, click **RUN QUERY** to see the traces that have been generated:

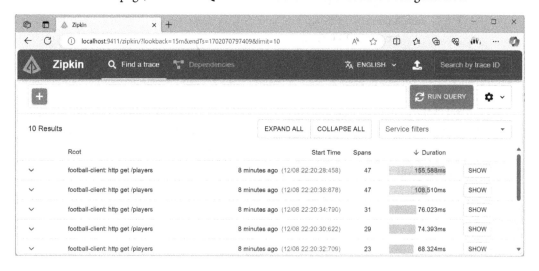

Figure 3.4: Root traces in Zipkin

On this page, you will see that the traces from the client application are root traces. Since we introduced a random error, you will see that there are failed and successful traces. If you click the **SHOW** button for any of these traces, you will see the traces of both RESTful APIs. There will be a main request for the client service and nested requests for the trading service:

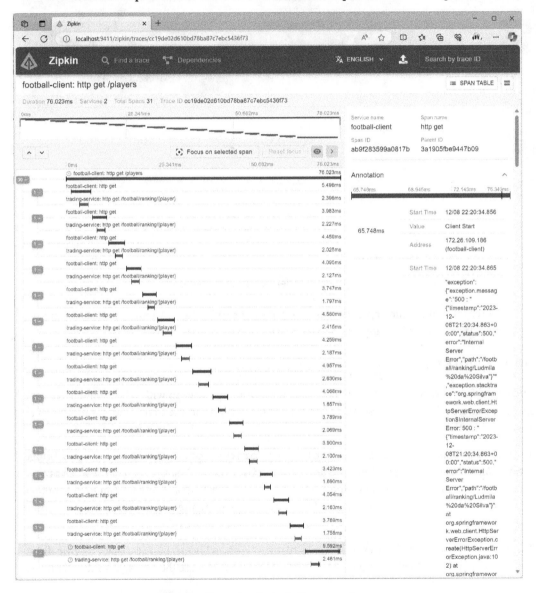

Figure 3.5: Trace details, including nested traces

You can also view the dependencies between services by clicking on the **Dependencies** link on the top bar:

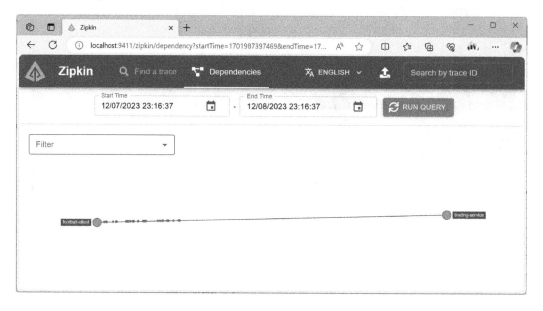

Figure 3.6: Viewing the dependencies between services in Zipkin

Here, you can see the dependencies between the `football-client` application and the `trading-service` application.

How it works...

Micrometer is a library that allows you to instrument your application without dependencies with specific vendors. This means that your code won't change if you decide to use another tool, such as *Wavefront*, instead of Zipkin.

The `io.micrometer:micrometer-tracing-bridge-otel` dependency creates a bridge between *Micrometer* and *OpenTelemetry*, after which the `io.opentelemetry: opentelemetry-exporter-zipkin` dependency exports from *OpenTelemetry* to *Zipkin*. If you want to use another tool to monitor your traces, you just need to change these dependencies, without any additional code changes.

The default address to send traces to Zipkin is `http://localhost:9411`. That's why we didn't need to configure it explicitly. In a production environment, you can use the `management.zipkin.tracing.endpoint` property.

In this recipe, we used `RestTemplateBuilder`. This is important as it configures `RestTemplate` by adding the tracing headers to the outgoing requests. Then, the target service gathers the tracing headers that can be used to nest the traces in the called application to the root trace from the client application. In reactive applications, you should use `WebClient.Builder` instead of `RestTemplateBuilder`.

In this recipe, we configured 100% sampling. This means that we send all traces to the tracing server. We did this for learning purposes; normally, you shouldn't do this in production as you can overload the tracing server by, for example, deploying a server via Zipkin or ingesting a lot of data if you're using a managed service in the cloud. The amount of data that's ingested directly affects monitoring systems – that is, the more data you ingest, the more it will cost you. However, even if you deploy your own tracing server, you will need to scale up as well. So, either way, it can increase your overall cost. In a large-scale system, having a sampling rate of 10% is more than enough to detect issues between services as well as understand the dependencies between the components.

There's more...

Micrometer tracing creates spans – that is, units of work or segments of a distributed trace that represent the execution of a specific operation, for each request. Spans capture information about the duration, context, and any associated metadata related to the respective operation.

You can create a span by starting an observation using the `ObservationRegistry` component. For instance, say `TradingService` has different important parts that you want to trace, such as *Collect data* and *Process data*. You can create different spans for those in your code.

To implement this, you will need to inject `ObservationRegistry` into your controller using the Spring Boot dependency container. For that, you need to define the `ObservationRegistry` parameter in the controller's constructor:

```
private final ObservationRegistry observationRegistry;
public FootballController(ObservationRegistry observationRegistry) {
        this.observationRegistry = observationRegistry;
}
```

Then, you must create the observations in the code:

```
@GetMapping("ranking/{player}")
public int getRanking(@PathVariable String player) {
    Observation collectObservation = Observation.
createNotStarted("collect", observationRegistry);
    collectObservation.lowCardinalityKeyValue("player", player);
    collectObservation.observe(() -> {
        try {
```

```
            logger.info("Simulate a data collection for player {}",
    player);
            Thread.sleep(random.nextInt(1000));
        } catch (InterruptedException e) {
            e.printStackTrace();
        }
    });

    Observation processObservation = Observation.
createNotStarted("process", observationRegistry);
    processObservation.lowCardinalityKeyValue("player", player);
    processObservation.observe(() -> {
            try {
                logger.info("Simulate a data processing for player
    {}", player);
                Thread.sleep(random.nextInt(1000));
            } catch (InterruptedException e) {
                e.printStackTrace();
            }
        });
        return random.nextInt(1000);
    }
```

Note that the observations include the player with `lowCardinalityKeyValue` to facilitate finding spans through this data.

Note

Some parts of the code have been removed for brevity. You can find the full version in this book's GitHub repository at `https://github.com/PacktPublishing/Spring-Boot-3.0-Cookbook/`.

Now, in Zipkin, you can see the custom spans nested in `trading-service`:

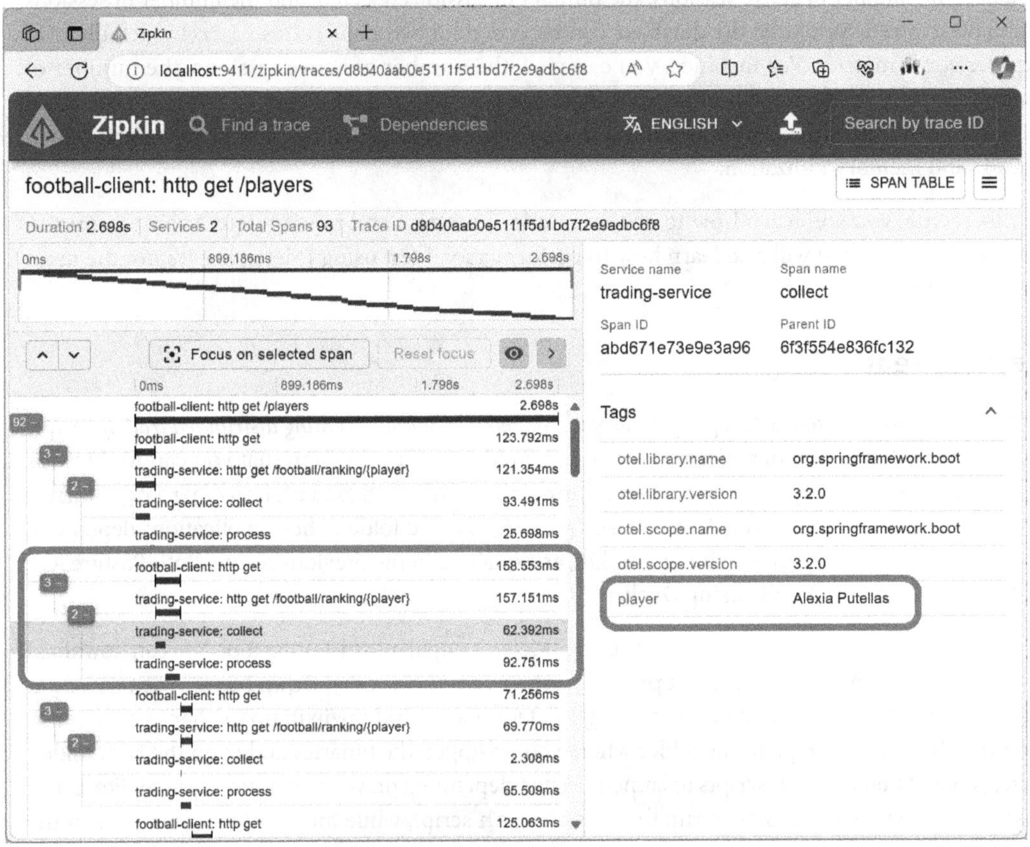

Figure 3.7: Custom spans in Zipkin

The `trading-service` span contains two nested spans, and both have a custom tag that specifies the player's name.

Accessing standard metrics

Your Football Trading service continues to grow by being adopted by football fans. You need to understand how it performs better so that you can adapt to demand while optimizing the resources that are used to provide the service.

You can use the standard metrics provided by Spring Boot Actuator and its related components for real-time insights into your application's behavior. For instance, you can find out how much CPU and memory has been used by your application or the time spent in **garbage collection (GC)**. These are the basic metrics that give you a general understanding of the performance of the application.

Other metrics are more subtle, such as the metrics provided by the web container, Tomcat – for instance, the number of active sessions, the number of sessions rejected, and the number of sessions that have expired. Similarly, the database connection pool, which is `hikaricp` by default, also exposes some metrics. For instance, you can view the number of active sessions, the number of waiting sessions, or the number of sessions that have been rejected. These types of metrics can be an indicator of problems in your application that aren't easy to detect just by using classic metrics such as CPU and memory utilization.

In this recipe, you will learn how to access standard metrics and how to detect some common application issues. You will also learn how to perform a load test using JMeter, but it's not the main purpose of this recipe.

Getting ready

In this recipe, you will reuse the applications you created in the *Implementing distributed tracing* recipe. If you haven't completed that recipe yet, I've prepared a working version that you can find in this book's GitHub repository at `https://github.com/PacktPublishing/Spring-Boot-3.0-Cookbook/`, in the `chapter3/recipe3-5/start` folder. These applications depend on PostgreSQL and also export activities to Zipkin, as explained in the previous recipe. Both PostgreSQL and Zipkin can be run locally using Docker.

In this recipe, we'll perform some load tests using JMeter, a popular load-testing tool. You can download JMeter from the project website at `https://jmeter.apache.org/download_jmeter.cgi`. Here, you can download a ZIP file containing JMeter binaries and unzip it; no further installation is required. To run JMeter, go to the folder where you unzipped the binaries and open the `bin` folder. Here, you will find different scripts to launch JMeter, depending on your operating system. For Unix-based operating systems, you can run the `jmeter.sh` script, while for Windows, you can run the `jmeter.bat` script.

I've created two JMeter scripts to create some load against the application. You can find them in this book's GitHub repository, in the `chapter3/recipe3-5/jmeter` folder.

How to do it...

In this recipe, we'll use the JMeter scripts mentioned in the *Getting ready* section to generate a workload for the football application. Then, we'll observe the metrics provided by Spring Boot and its related components. Follow these steps:

1. Before running the first load test, ensure that the trading application is running and the `metrics` endpoint is exposed. As explained in the *Adding Actuator to your application* recipe, this can be done by adding the `metrics` value to the `management.endpoints.web.exposure.include` parameter. If you followed the previous recipes or used the working version I've prepared, as explained in the *Getting ready* section, the `application.yml` file should look like this:

    ```
    management:
        endpoints:
            web:
                exposure:
                    include:
    health,env,metrics,beans,loggers,football
    ```

2. Once the application is running, open JMeter and load the `loadTeams.jmx` script. You can find it in the `chapter3/recipe3-5/jmeter` folder, as explained in the *Getting ready* section. This script makes a request to the application's `/football` path and returns a list of teams. This process is executed by 30 threads infinitely.

    You can adjust some parameters of the load tests depending on the resources of your development computer. For instance, I used 30 threads to overload my computer, but maybe you need more or even fewer threads than that:

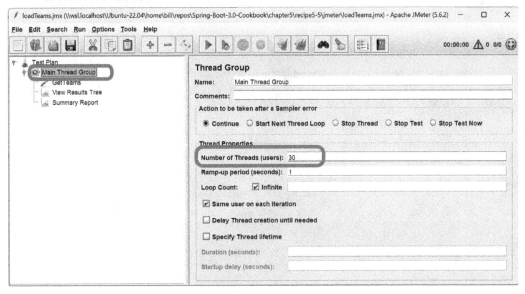

Figure 3.8: The number of threads in JMeter

If you want to adjust the number of threads, click **Main Thread Group** and adjust **Number of Threads (users)**.

Once the application is ready, you can run the JMeter script.

3. Let's observe the metrics of the application. Go to `http://localhost:8080/actuator/metrics` to see the full list of exposed metrics. You can get any of these metrics by appending the metric's name to the `/actuator/metrics` path. Typically, you will get the CPU and memory-related counters:

    - With `http://localhost:8080/actuator/metrics/process.cpu.usage`, you will get the percentage of CPU being used by the application process

    - With `http://localhost:8080/actuator/metrics/system.cpu.usage`, you will get the percentage of CPU being used by the system

    - With `http://localhost:8080/actuator/metrics/jvm.memory.used` you will get the amount of memory being used by your application

    As an example, the result of the `process.cpu.usage` metric looks like this:

    ```
    {
        "name": "system.cpu.usage",
        "description": "The \"recent cpu usage\" of the system the
    application is running in",
        "measurements": [
            {
                "statistic": "VALUE",
                "value": 0.48494983277591974
            }
        ],
        "availableTags": []
    }
    ```

4. Stop the test – you need to create a new endpoint to access the database. For that, follow these steps:

    I.   Create a new `DataService` class and inject `JdbcTemplate` into the constructor:

    ```
    @Service
    public class DataService {
        private JdbcTemplate jdbcTemplate;
        public DataService(JdbcTemplate jdbcTemplate) {
            this.jdbcTemplate = jdbcTemplate;
        }
    }
    ```

II.   Now, create a method that makes a call to the database. To simulate a slow database query, you can use the `pg_sleep` PostgreSQL command. This command waits for a given number of seconds or fraction of seconds:

```
public String getPlayerStats(String player) {
    Random random = new Random();
    jdbcTemplate.execute("SELECT pg_sleep(" + random.
nextDouble(1.0) + ")");
    return "some complex stats for player " + player;
}
```

Since we passed a random number between 0 and 1.0, it will wait a fraction of a second in the database.

III.   You can inject the new `DataService` into the `FootballController` class and create a method that will use the new service:

```
@GetMapping("/stats/{player}")
public String getPlayerStats(@PathVariable String player){
    return dataService.getPlayerStats(player);
}
```

Now, you can run the application.

5.   Finally, run another JMeter script that makes a request to the same /football path and returns a list of teams, as well as the new path, /stats/{player}, which performs a long request to the database. Again, 30 threads are running these requests infinitely.

How it works...

In the first load test, we can see that there is a bottleneck in the application's CPU. In a real-world scenario, the CPU metric can be used to scale the application automatically, such as by adding new instances of the application. That's the kind of bottleneck that we could expect under heavy loads.

In the second load test, there is no physical resource bottleneck, but there's a query that takes a long time and blocks a connection that cannot be reused for other requests. In a real-world scenario, you could increase the number of available connections in the connection pool, but only up to a certain limit, since this is a very expensive and finite resource.

If you look at `system.cpu.usage` and `process.cpu.usage`, you will see that the values are much lower than 1.0, which we observed in the previous load test.

You can also look at the metrics related to the database connection pool. The default database connection pool in Spring Data is HikariCP, and all the metrics related to this component are `hikaricp.*`. Let's consider the following metrics:

- `hikaricp.connections.max`: This value specifies the maximum number of real database connections that `hikaricp` will open in the PostgreSQL server. This number won't change during the execution of the test as the value is static during the application life cycle. By default, it's set to 10.

- `hikaricp.connections.active`: This is the number of active connections – that is, the connections that are executing something in the database server. Under light loads, the number will be less than the maximum. Since the database operation is long (up to 1 second), and there are 30 concurrent threads for only 10 maximum connections, this number will be 10 or near 10 during the execution of the JMeter script.

- `hikaricp.connections.pending`: When there are no available connections in the connection pool, this metric queues the requests. This metric specifies the number of connections waiting for an available connection. This number will be greater than 1 during the JMeter script's execution.

- `hikaricp.connections.timeout`: If a request is waiting for more than a given amount of time –30 seconds by default – it will time out. After executing the JMeter script, you will see that this metric will be more than 1.

Opening a physical database connection is an expensive operation. To avoid the overhead of creating a connection, there is a mechanism known as a database connection pool that keeps some already created connections ready to be used. When a process needs to connect to the database, it gets the connection from the pool and returns it to the pool once the operation is finished. In the second stress test, there were no connections as they took a long time to complete, so they took a long time to return to the pool. When there are no available connections, the connection pool enqueues the connection until one is released. That's why you saw `pending` connections. After some time, you will see timeout connections. Those are the connections that were enqueued for more than 30 seconds.

This situation also impacts the web container. By default, the number of threads to serve HTTP requests is finite and there is also a pool. When there are no more available threads, the web container – in this case, Tomcat – will enqueue the requests. In this kind of situation, when an HTTP request is mostly waiting for a dependency to complete, it appears the Reactive framework. In this case, the application uses special kinds of threads – non-blocking threads – that are intended for I/O operations. These types of threads allow the application to continue processing other tasks while waiting for responses from external services.

See also

You can visualize your metrics with standard monitoring tools. In the *Integrating your application with Prometheus and Grafana* recipe, you will learn how to integrate application metrics with Prometheus and visualize them with Grafana. These are two popular open source tools that are part of the **Cloud Native Computing Foundation (CNCF)**.

Creating your own metrics

So far, you've created a new feature in your Football Trading service where users can list a card for exchange and another user can bid for the traded card. When a new bid is received, it is queued in memory until it is committed as it requires a bunch of complex validations. There are a lot of expectations for this new feature, and you want to be sure it works well. For that reason, you want to monitor the bids that are received, how many bids are pending to be committed, and the duration of this process.

In this recipe, you will learn how to create custom metrics using **Micrometer**. Micrometer is an open source metrics collection library for Java applications that is very well integrated with Spring Boot Actuator. Other libraries can use the telemetry data generated by Micrometer to export to different monitoring systems.

There are different types of metrics:

- **Counter**: As the name suggests, it counts how many times something happened. We can use this type of metric to find out how many bids were received.

- **Gauge**: This metric provides a value in a given moment. We can use it to find out how many bids are waiting to be processed.

- **Timer**: This metric measures the duration of a given operation. We can use it to find out the time spent per bid.

Getting ready

In this recipe, we'll reuse the projects from the *Accessing standard metrics* recipe. I've prepared a working version if you haven't completed that recipe yet. You can find it in this book's GitHub repository at `https://github.com/PacktPublishing/Spring-Boot-3.0-Cookbook/`, in the `chapter3/recipe3-6/start` folder.

To simulate a workload for the new feature, I've created a JMeter script. You can find it in this book's GitHub repository, in the `chapter3/recipe3-6/jmeter` folder. You can download JMeter from the project website at `https://jmeter.apache.org/download_jmeter.cgi`. Here, you can download a ZIP file that contains JMeter binaries and unzip it – no further installation is required. To run JMeter, go to the folder where you unzipped the binaries, then open the `bin` folder. Here, you can find different scripts to launch JMeter, depending on your operating system. For Unix, you can run the `jmeter.sh` script, while for Windows, you can run the `jmeter.bat` script.

How to do it...

In this recipe, you'll incorporate your custom metrics into the football trading application. This enhancement will offer improved insights into your application's performance during runtime:

1.   Go to your trading application and create a new service class named `AuctionService`:

    I.   Inject `MeterRegistry` into the constructor. In the same constructor, create a counter for the bids received, a timer for the duration of the bid to be processed, and a gauge for the bids waiting to be confirmed. The class should look like this:

```
@Service
public class AuctionService {
    private Map<String, String> bids = new
ConcurrentHashMap<>();
    private Counter bidReceivedCounter;
    private Timer bidDuration;
    Random random = new Random();
    public AuctionService(MeterRegistry meterRegistry) {
        meterRegistry.gauge("football.bids.pending", bids,
Map::size);
        this.bidReceivedCounter = meterRegistry.
counter("football.bids.received");
        this.bidDuration = meterRegistry.timer("football.bids.
duration");
    }
}
```

    II.   Note that `gauge` returns the size of the map that's used to keep the bids that have been received in memory.

    III. Now, create a method to process the bids. In this method, you will use the `bidDuration` timer to measure the duration of the operation and increase the number of bids received using `bidReceivedCounter`.

    IV.   Use the `ordersTradedCounter` and `tradedDuration` metrics in a new method named `tradeCards`. The method should look like this:

```
public void addBid(String player, String bid) {
    bidDuration.record(() -> {
        bids.put(player, bid);
        bidReceivedCounter.increment();
        try {
            Thread.sleep(random.nextInt(20));
        } catch (InterruptedException e) {
            e.printStackTrace();
        }
```

```
        bids.remove(player);
   });
}
```

2. Next, expose this feature in the `FootballController` class:

    I.   Inject your new `AuctionService` into the constructor:

    ```
    private AuctionService auctionService;
    public FootballController(AuctionService auctionService) {
       this.auctionService = auctionService;
    }
    ```

    II. Note that all the other parameters and fields have been omitted for simplicity. Since we are reusing the same project from previous recipes, you should have more parameters in the constructor, and you should have other fields as well.

    III. Create a new method that will present bids for players using the new service:

    ```
    @PostMapping("/bid/{player}")
    public void addBid(@PathVariable String player,
                                @RequestBody String bid) {
          auctionService.addBidAOP(player, bid);
    }
    ```

3. Now, you can run the application and start generating some load. To do this, open the `loadBids.jmx` file in JMeter. You can find this file in this book's GitHub repository at `https://github.com/PacktPublishing/Spring-Boot-3.0-Cookbook/`, in the `chapter3/recipe3-6/jmeter` folder. Then, run the script in JMeter and keep it running while you observe the metrics.

4. Observe the counters you created:

   • If you open the Actuator metrics endpoint at `http://localhost:8080/actuator/metrics`, you will see the new metrics that have been created: `football.bids.duration`, `football.bids.pending`, and `football.bids.receieved`. If you append the names of these metrics to the Actuator metrics endpoint, you will get the values of each metric.

   • Open `http://localhost:8080/actuator/metrics/football.bids.received` to get the number of bids that have been received. You will see the total number of bids.

   • Open `http://localhost:8080/actuator/metrics/football.bids.duration` to get the bids processing duration.

   • Open `http://localhost:8080/actuator/metrics/football.bids.pending` to get the number of bids that are pending.

For counters and duration, normally, the monitoring tools also provide a rate that's calculated from the total values and based on the frequency of observation. It's more interesting in terms of performance analysis to know the bid processing rate than the total number. The same goes for the duration.

5. Stop the JMeter script.

How it works...

The `MeterRegistry` class registers the metrics, after which they are automatically exposed in the Actuator metrics endpoint.

`gauge` calls the delegate that's been assigned to the metric. This delegate will be executed according to the observation frequency. In this recipe, we call the endpoint explicitly. If you use a monitoring tool, it will be observed periodically. Keep in mind that this operation should be as lightweight as possible because it will be called frequently.

Timer metrics measure the time spent on the execution of the delegate provided.

A counter metric increments the value of the counter. If you don't provide a value when calling the `increment` method, as we did in this recipe, it just increments by 1. You can provide a number as a parameter of method increment, at which point it will increment the counter value by the number provided. This number should always be positive.

There's more...

You can create metrics by using a more declarative approach with **Aspect Oriented Programming** (**AOP**) libraries. For this, you should add a dependency to the *AOP starter* and configure the `ObservedAspect` bean.

To add the dependency to the *AOP starter*, include the following in your pom.xml file:

```
<dependency>
    <groupId>org.springframework.boot</groupId>
    <artifactId>spring-boot-starter-aop</artifactId>
</dependency>
```

To configure the `ObserverAspect` bean, add the following method to the `Football Configuration` class:

```
@Bean
ObservedAspect observedAspect(ObservationRegistry observationRegistry)
{
    return new ObservedAspect(observationRegistry);
}
```

At this point, you can use the @Observed annotation in your code to generate metrics automatically. For instance, in this recipe, we could annotate the AuctionService class with @Observed:

```
@Observed(name = "football.auction")
@Service
public class AuctionService {
}
```

Then, you can simplify the class as you don't need to explicitly create the counters in the constructor. In the addBidAOP method, you only need to focus on the application logic:

```
public void addBidAOP(String player, String bid) {
    bids.put(bid, player);
    try {
        Thread.sleep(random.nextInt(20));
    } catch (InterruptedException e) {
        e.printStackTrace();
    }
    bids.remove(bid);
}
```

When you run the application and AuctionService is used (the metrics are created lazily the first time the methods are used), you will see that there are two new metrics in the Actuator metrics endpoint:

- football.auction: Provides general counters for the methods defined in your annotated class
- football.auction.active: Provides counters for active executions for the methods defined in your annotated class

The following is a sample of the football.auction metric that was obtained from http://localhost:8080/actuator/endpoint/football.auction:

```
{
    "name": "football.auction",
    "baseUnit": "seconds",
    "measurements": [
        {
            "statistic": "COUNT",
            "value": 1648870
        },
        {
            "statistic": "TOTAL_TIME",
            "value": 15809.168264051
        },
        {
```

```
            "statistic": "MAX",
            "value": 0.02272261
        }
    ],
    "availableTags": [
        {
            "tag": "method",
            "values": [
                "addBidAOP"
            ]
        },
        {
            "tag": "error",
            "values": [
                "none"
            ]
        },
        {
            "tag": "class",
            "values": [
                "com.packt.footballobs.service.AuctionService"
            ]
        }
    ]
}
```

You can get metrics for a specific method using tags. For instance, to get the metrics of the addBidAOP method, you can perform the following request: http://localhost:8080/actuator/metrics/football.auction?tag=method:addBidAOP.

This service is implemented in this book's GitHub repository at https://github.com/PacktPublishing/Spring-Boot-3.0-Cookbook, in the chapter3/recipe3-8/end folder. As mentioned previously, the metric is created lazily, so you should invoke this service to make it available. You can do this by executing the following curl request in your terminal:

```
curl http://localhost:8080/football/bid/357669 \
--request POST \
--data "200"
```

Integrating your application with Prometheus and Grafana

You have a successful Football Trading application, and you can observe it by calling the various Actuator endpoints. However, this way of observing the application is too manual. So, you want a system that allows you to automate how your application is monitored.

In this recipe, you will learn how to expose the metrics of your application using a format that can be used by **Prometheus**, after which you will use the Prometheus data as a source for **Grafana**. Prometheus is an open source monitoring solution that collects and aggregates metrics as time series data, then stores the events in real time so that the events can be used to monitor your application. Grafana is an open source tool for visualization that allows you to create custom dashboards, graphs, and even alerts. One of the sources Grafana can use is the data collected by Prometheus. The combination of both tools is a very popular choice due to its ease of use, flexibility, and scalability.

Getting ready

In this recipe, you will reuse the outcome of the *Creating your own metrics* recipe. I've prepared a working version of this in case you haven't completed it yet. You can find it in this book's GitHub repository at `https://github.com/PacktPublishing/Spring-Boot-3.0-Cookbook/`, in the `chapter3/recipe3-7/start` folder.

You will use Prometheus and Grafana servers. As usual, the easiest way to run Prometheus and Grafana on your local computer is by using Docker.

To download and start Prometheus, run the following command in your terminal:

```
docker run -d --name prometheus -p 9090:9090 \
-v prometheus.yml:/etc/prometheus/prometheus.yml \
prom/prometheus
```

This command uses the `-v` parameter to mount a volume to a file named `prometheus.yml`. This file contains the configuration for Prometheus. The configuration will be described and created as part of this recipe in the *How to do it...* section.

To download and start Grafana, run the following command in your terminal:

```
docker run -d --name grafana -p 3000:3000 grafana/grafana
```

To simulate a workload for the new feature, I've created a JMeter script. You can find it in this book's GitHub repository, in the `chapter3/recipe3-7/jmeter` folder. You can download JMeter from the project's website at `https://jmeter.apache.org/download_jmeter.cgi`. From here, download a ZIP file containing JMeter binaries and unzip it; no further installation is required. To run JMeter, go to the folder where you unzipped the binaries, then open the `bin` folder. Here, you will find different scripts to launch JMeter, depending on your operating system. For Unix, you can run the `jmeter` script, while for Windows, you can run the `jmeter.bat` script.

How to do it...

First, we'll configure our application so that it exposes a Prometheus endpoint. Afterward, we'll set up Prometheus and Grafana so that we can ingest the data provided by our application:

1. Let's start by exposing a Prometheus endpoint to the trading application. For that, two steps are necessary:

    I.   Add the following dependency to the `pom.xml` file:

    ```
    <dependency>
        <groupId>io.micrometer</groupId>
        <artifactId>micrometer-registry-prometheus</artifactId>
    </dependency>
    ```

    II.   Expose the Prometheus endpoint. To do so, open the `application.yml` file in the `resources` folder and add the following highlighted properties:

    ```
    management:
        endpoint:
            health:
                probes:
                    enabled: true
            prometheus:
                enabled: true
        endpoints:
            web:
                exposure:
                    include:
    health,env,metrics,beans,loggers,football,prometheus
    ```

2. You can run the application and open the Prometheus endpoint at `http://localhost:8080/actuator/prometheus`.

3.   The next step is running Prometheus and configuring it to consume the newly exposed endpoint. You can configure Prometheus by creating a `.yaml` configuration file and mounting it on the Prometheus Docker image:

    I.   Prometheus will be hosted on Docker, while the application will be hosted on your computer, the Docker host. The first task is obtaining the IP address of your computer. On Linux, you can run the following command in your Terminal:

```
ip addr show
```

    II.   On Windows, you can run the following command in your terminal:

```
ipconfig
```

    III.   If you run your application in **Windows Subsystem for Linux (WSL)**, you should run `ip addr show` in a WSL terminal.

    IV.   For instance, the IP of my interface is `172.26.109.186` when I run `ip addr show`. I will use this value to configure the Prometheus YAML file.

    V.   Let's continue by creating the configuration file using the IP address we obtained in the previous step. In the project's root directory, create an application named `prometheus.yml` with the following content:

```
global:
  scrape_interval: 3s
scrape_configs:
  - job_name: 'football_trading_app'
    metrics_path: '/actuator/prometheus'
    static_configs:
      - targets: ['172.26.109.186:8080']
```

    VI.   Note that we configured the metrics path exposed by our application, and the target is the IP address and port of our application.

    VII.   Now, run the Prometheus container using the configuration file. For that, in the same directory you created the configuration file, execute the following command in your terminal:

```
docker run -d --name prometheus -p 9090:9090 \
 -v $(pwd)/prometheus.yml:/etc/prometheus/prometheus.yml prom/
prometheus
```

    VIII.   This command runs the `prom/prometheus` image, exposing port `9090` and mounting the `prometheus.yml` file in the container filesystem at `/etc/prometheus/prometheus.yml`. `$(pwd)` is a command substitution in Linux that is used to insert the current directory.

4.   Now, Prometheus should be working and *scrapping* your application to get observability data. To verify it's working, you can open Prometheus at `http://localhost:9090`, then open the **Status** menu and select **Targets**:

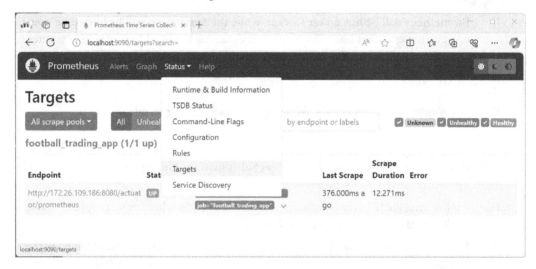

Figure 3.9: Prometheus targets

Verify that the status of your target is working. It should be **UP**.

5.   You can use Prometheus to visualize the data from your application. Go to the Prometheus home page, search for any metric, and click on **Execute** to see the data. If you select the **Graph** tab, you will see the data in graphical form:

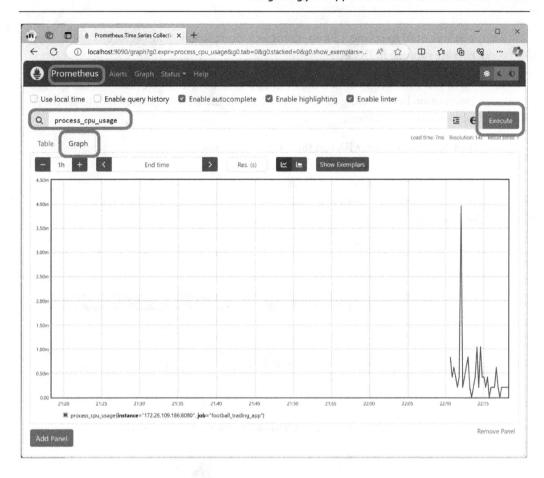

Figure 3.10: Visualizing data in Prometheus

6.   The visualization capabilities that are available in Prometheus are a bit limited, but we can use Grafana and connect it to Prometheus to achieve better visualization:

I.   Ensure that Grafana is running. As explained in the *Getting ready* section, you can run Grafana using Docker by executing the following command in your terminal:

```
docker run -d --name grafana -p 3000:3000 grafana/grafana
```

II.   Now, you can open Grafana by opening the following address in your browser: `http://localhost:3000`. You will be asked for your credentials. You can use the default credentials – that is, user set to `admin` and password set to `admin`.

7. Next, you will need to connect Prometheus as a Grafana data source. At this point, both containers are running in Docker:

   I. First, you will need to obtain the Prometheus IP address in Docker. You can get this information by inspecting the container. Execute the following commands to get the IP address of the container:

- To retrieve the container ID, run the following command:

```
docker ps
```

- My container ID was 5affa2883c43. Replace this with your container ID when running the following command:

```
docker inspect 5affa2883c43 | grep IPAddress
```

My terminal looks like this:

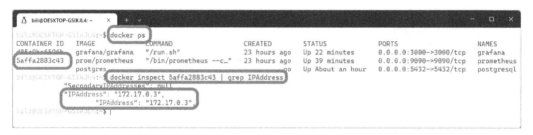

Figure 3.11: Using docker inspect to get the container's IP address

II.   Now, open the menu on the left and select **Connections | Data sources**:

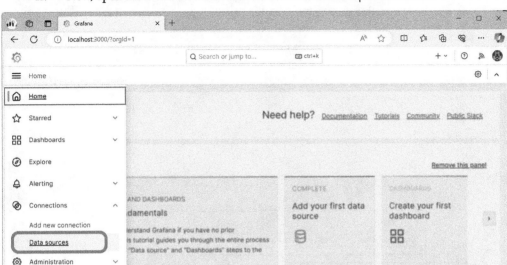

Figure 3.12: Opening Data sources

Click **Add data source** and select **Prometheus** as a data source type. If it doesn't appear on the first page, search for Prometheus in the search bar:

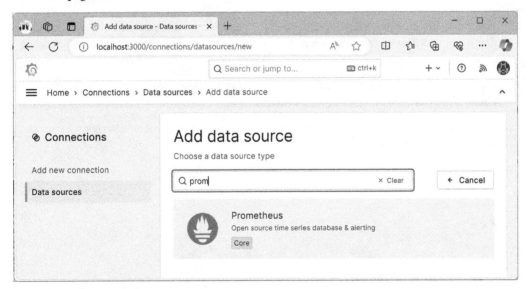

Figure 3.13: Selecting Prometheus as a data source

Then, configure the **Prometheus server URL** property. You will need to use the IP you obtained previously. In my case, this is 172.17.0.3, but you likely have another value. The port is 9090:

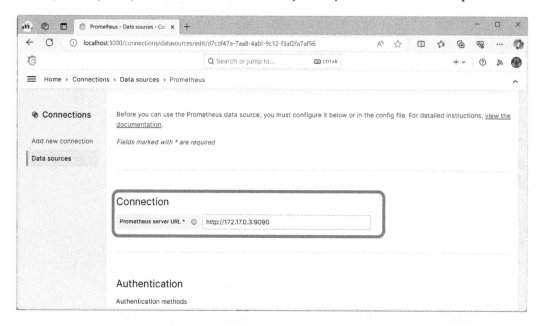

Figure 3.14: Configuring the Prometheus server URL property

You can keep the default value for the rest of the parameters. At the bottom of the page, you'll find the **Save & Test** button. Click on it. At this point, you can start visualizing data by building a dashboard.

8. Finally, create a dashboard to visualize the number of pending bids. Go to **Dashboards**, click **Create Dasboard**, and then click **Add visualization**.

For **Metric**, select football_bids_pending, and then click **Run queries**. Change the time range to the last 30 minutes. Finally, click **Save**:

Figure 3.15: Configuring a panel

Now, save your dashboard. Name it `Pending Bids`.

9. Run a load test to see how metrics are visualized in the panel. You can use the JMeter script I created to generate some traffic. You can find it in this book's GitHub repository at https://github.com/PacktPublishing/Spring-Boot-3.0-Cookbook/, in the chapter3/recipe3-7/jmeter folder. The Grafana panel should look like this:

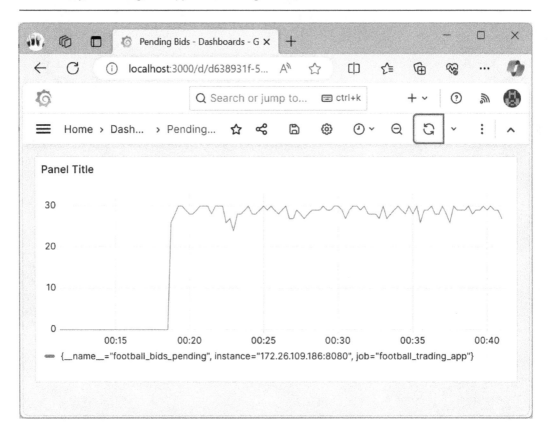

Figure 3.16: Pending bids visualized in Grafana

With that, you've learned how to visualize your metrics in powerful tools such as Grafana.

How it works...

Prometheus is an extensible tool that can use exporters. These exporters are jobs that run in Prometheus and can get data from external sources if they're exposed using the appropriate format. This recipe's job scrapes the data, meaning that it gets the data from the external source periodically. In this recipe, we configured our application to export the data in a format that Prometheus can understand, after which we configured a target to retrieve that data.

Some of the benefits of using Prometheus are as follows:

- It can take metrics from multiple sources – not only applications but also infrastructure components.
- It allows PromQL to be used, a language for querying and aggregating data. You can combine this data from multiple sources to extract relevant information for monitoring.

- You can create alerts based on queries and the thresholds you define. For instance, we could use CPU usage thresholds or our pending bids to send an alert.

Grafana can take data from different sources; one of them is Prometheus. This combination is very popular for monitoring solutions. Grafana can be used for advanced visualizations, and it also allows you to create alerts and send notifications. This is very important as it improves the monitoring automation process.

In this recipe, we used these popular open source tools, but the same approach can be used with other commercial tools. Usually, monitoring tools manage tracing, logging, and metrics, adding capabilities for visualization, such as dashboards and alerting by different channels.

An important thing to think about is when you should use traces or metrics for monitoring. Traces are very useful in showing the relationship between services and finding the specific operations using data from the transaction itself. This is very helpful in finding the root cause of an issue. The main issue with traces is that in scenarios with a high volume of operations, the amount of data that's generated can be huge, and usually, the traces are sampled so that all the data that's been generated can be processed and the cost can be controlled.

On the other hand, the metrics aggregate the measurements, and they just export those aggregated measurements periodically to create the time series data. Then, the data that's generated is constant, regardless of the traffic managed by the target system. The main advantage of metrics is that they don't require sampling and the data that's generated is quite precise. For that reason, the metrics are more appropriate for certain types of alerts. However, when you need to find the root cause of an issue, traces are more appropriate.

Changing the settings of a running application

So far, you've added logging to your successful football trading application, and it receives quite a lot of traffic. The program creates logs in different places. These logs can help you figure out what the program did while it was running. Not every log is equally important. So, the program uses various log levels, ranging from debugging to error logs. Sorting logs by their level prevents an excessive number of logs from being created. However, you want to ensure you can change the minimum level of logs to be processed without restarting or redeploying your application.

Some Spring Boot Actuator endpoints allow you to make changes in runtime, with no need to restart the application. The logging endpoint is one of those endpoints as it allows you to change the minimum level of logging.

In this recipe, you will learn how to change the logging level of a running application.

Getting ready

In this recipe, you will reuse the outcome of the *Integrating your application with Prometheus and Grafana* recipe. I've prepared a working version in case you haven't completed it yet. You can find it in this book's GitHub repository at `https://github.com/PacktPublishing/Spring-Boot-3.0-Cookbook/`, in the `chapter3/recipe3-8/start` folder.

How to do it...

In this recipe, you'll adapt the football trading application so that it generates logs with different levels of importance. Once you've done this you'll learn how to change the level at runtime:

1. First, let's add some logs to the `TradingService` class:

    I.   Create a logger for the class. You can define a static member for this purpose:

    ```
    private static final Logger logger = LoggerFactory.
    getLogger(TradingService.class);
    ```

    II.   Then, add debug and information logging to the `getPendingOrders` method:

    ```
    public int getPendingOrders() {
        logger.debug("Ensuring that pending orders can be
    calculated");
        Random random = new Random();
        int pendingOrders = random.nextInt(100);
        logger.info(pendingOrders + " pending orders found");
        return pendingOrders;
    }
    ```

    III. You can also add some logging for the `tradeCards` method:

    ```
    public int tradeCards(int orders) {
        if (getPendingOrders() > 90) {
            logger.warn("There are more than 90 orders, this can
    cause the system to crash");
            AvailabilityChangeEvent.
    publish(applicationEventPublisher, new Exception("There are more
    than 90 pending orders"), LivenessState.BROKEN);
        } else {
            logger.debug("There are more less than 90 orders, can
    manage it");
            AvailabilityChangeEvent.
    publish(applicationEventPublisher, new Exception("working
    fine"), LivenessState.CORRECT);
        }
        return orders;
    }
    ```

2. Now, you can perform some requests and validate that the information is being logged. You can execute the following command in your terminal to execute a request every second:

```
watch -n 1 -x curl --request POST -H "Content-Type: application/
json" --data "1" http://localhost:8080/football
```

You will see that only `INFO` and `WARN` logs are processed:

Figure 3.17: Only INFO and WARN logs are processed

This is because the default level is `INFO`. This means that only `INFO` or higher priority levels are logged.

3. You can verify the log level by calling the Actuator `loggers` endpoint. Go to `http://localhost:8080/actuator/loggers`. You will see the available log levels, as well as the loggers that are defined in your application. You will see that there is a logger for your service class, `com.packt.footballobs.service.TradingService`, and that the effective level is `INFO`.

4. Let's say you've detected an issue in the application, and you want to activate the `DEBUG` level. Let's change it by using the Actuator `loggers` endpoint. For that, you just need to perform the following request:

```
curl --request POST \
-H 'Content-Type: application/json' \
-d '{"configuredLevel": "DEBUG"}' \
http://localhost:8080/actuator/loggers/com.packt.footballobs.
service.TradingService
```

You will see that it now generates logs for `DEBUG` as well:

```
2024-06-04T00:04:33.096+02:00 DEBUG 40677 --- [trading-service] [nio-8080-exec-5] [906e95eb3f92032bb9ad58c8a7a0f7ec-e22b575f9d3a8154] c.p.footballobs.service.TradingService   : There are more less th
an 90 orders, can manage it
2024-06-04T00:04:34.109+02:00 DEBUG 40677 --- [trading-service] [nio-8080-exec-7] [5925e5fb7d52921d104b30f4fd6f357c-4fc0e9e4c9351e13] c.p.footballobs.service.TradingService   : Ensuring that pending
orders can be calculated
2024-06-04T00:04:34.110+02:00  INFO 40677 --- [trading-service] [nio-8080-exec-7] [5925e5fb7d52921d104b30f4fd6f357c-4fc0e9e4c9351e13] c.p.footballobs.service.TradingService   : 37 pending orders foun
d
2024-06-04T00:04:34.110+02:00 DEBUG 40677 --- [trading-service] [nio-8080-exec-7] [5925e5fb7d52921d104b30f4fd6f357c-4fc0e9e4c9351e13] c.p.footballobs.service.TradingService   : There are more less th
an 90 orders, can manage it
2024-06-04T00:04:35.119+02:00 DEBUG 40677 --- [trading-service] [nio-8080-exec-9] [cd21529a8c75be873c5173852b5ad628-827ccdff970efc86] c.p.footballobs.service.TradingService   : Ensuring that pending
orders can be calculated
2024-06-04T00:04:35.119+02:00  INFO 40677 --- [trading-service] [nio-8080-exec-9] [cd21529a8c75be873c5173852b5ad628-827ccdff970efc86] c.p.footballobs.service.TradingService   : 5 pending orders found
2024-06-04T00:04:35.119+02:00 DEBUG 40677 --- [trading-service] [nio-8080-exec-9] [cd21529a8c75be873c5173852b5ad628-827ccdff970efc86] c.p.footballobs.service.TradingService   : There are more less th
an 90 orders, can manage it
2024-06-04T00:04:36.128+02:00 DEBUG 40677 --- [trading-service] [nio-8080-exec-1] [f436eb112a8d85db88eb388411c66ff0-1883e8440efd36b1] c.p.footballobs.service.TradingService   : Ensuring that pending
orders can be calculated
2024-06-04T00:04:36.128+02:00  INFO 40677 --- [trading-service] [nio-8080-exec-1] [f436eb112a8d85db88eb388411c66ff0-1883e8440efd36b1] c.p.footballobs.service.TradingService   : 34 pending orders foun
d
2024-06-04T00:04:36.128+02:00 DEBUG 40677 --- [trading-service] [nio-8080-exec-1] [f436eb112a8d85db88eb388411c66ff0-1883e8440efd36b1] c.p.footballobs.service.TradingService   : There are more less th
an 90 orders, can manage it
2024-06-04T00:04:37.138+02:00 DEBUG 40677 --- [trading-service] [nio-8080-exec-3] [76db460ebc3cf324b2174078cf4569ed-03fcaecb8250da88] c.p.footballobs.service.TradingService   : Ensuring that pending
orders can be calculated
2024-06-04T00:04:37.138+02:00  INFO 40677 --- [trading-service] [nio-8080-exec-3] [76db460ebc3cf324b2174078cf4569ed-03fcaecb8250da88] c.p.footballobs.service.TradingService   : 92 pending orders foun
d
2024-06-04T00:04:37.138+02:00  WARN 40677 --- [trading-service] [nio-8080-exec-3] [76db460ebc3cf324b2174078cf4569ed-03fcaecb8250da88] c.p.footballobs.service.TradingService   : There are more than 90
 orders, this can cause the system to crash
2024-06-04T00:04:38.146+02:00 DEBUG 40677 --- [trading-service] [nio-8080-exec-5] [0492e5d015aeec856a2f8d525c9c9fc2-79d936f1a946dc57] c.p.footballobs.service.TradingService   : Ensuring that pending
orders can be calculated
2024-06-04T00:04:38.146+02:00  INFO 40677 --- [trading-service] [nio-8080-exec-5] [0492e5d015aeec856a2f8d525c9c9fc2-79d936f1a946dc57] c.p.footballobs.service.TradingService   : 68 pending orders foun
d
2024-06-04T00:04:38.146+02:00 DEBUG 40677 --- [trading-service] [nio-8080-exec-5] [0492e5d015aeec856a2f8d525c9c9fc2-79d936f1a946dc57] c.p.footballobs.service.TradingService   : There are more less th
an 90 orders, can manage it
```

Figure 3.18: DEBUG and higher critical logs are generated

If you verify the `loggers` endpoint, as explained in *Step 3*, you will see that the `TradingService` class now has two attributes:

- `configuredLevel`: `DEBUG`
- `effectiveLevel`: `DEBUG`

5. Now that you've verified the logs, you decide to change the log level to `WARN` by running the following command since too much noise is generated by `DEBUG` and `INFO` logs:

```
curl --request POST \
-H 'Content-Type: application/json' \
-d '{"configuredLevel": "WARN"}' \
http://localhost:8080/actuator/loggers/com.packt.footballobs.
service.TradingService
```

If you verify the `loggers` endpoint, as explained in *Step 3*, you will see that the `TradingService` level is `WARN`. If you continue making requests, you will see that only `WARN` logs are emitted.

How it works...

As we saw in the *Creating a custom Actuator endpoint* recipe, some endpoints implement update and delete operations. The `loggers` endpoint allows you to change the log level. This is a very helpful feature when you need to find issues in production as you no longer need to restart your application.

In an application with high traffic, you will usually want to have a high log level, such as WARN. This is the warning level and is typically used to indicate that there is a potential issue or anomaly that should be noted. It signifies a situation that may not necessarily be an error, but it could lead to problems if it's not addressed. The reason for using higher log levels, such as WARN, is that the logs are usually saved by the monitoring system. If the application generates too many logs, it requires more resources to process and retain them, and that can be costly. At the same time, DEBUG and INFO logs are not critical and they can generate too much information, making it more difficult to find the root cause of the problems.

There's more...

Other standard endpoints are part of Spring Boot that allow you to make changes at runtime. For instance, the sessions endpoint allows you to retrieve and delete user sessions.

4

Spring Cloud

In modern systems, you may find several microservices interacting with each other. Spring Cloud offers easy-to-deploy components that simplify the interaction and coordination of distributed systems to tackle large-scale application challenges such as scalability, availability, observability, and resilience.

In this chapter, you will learn how to use Spring Cloud components to develop a scalable and resilient distributed system. You will build upon the learnings of the previous chapters and configure security and observability in this Spring Cloud setup. This will help you effectively monitor and troubleshoot your distributed architecture. By the end of this journey, you will know how to design and develop cloud-native applications.

Finally, you will be taught how to deploy Spring Boot Admin, a widely used open source project in the Spring ecosystem. This project offers a user-friendly web interface that enables you to monitor and manage multiple Spring Boot applications centrally. Additionally, it can be effortlessly integrated with other Spring Cloud components.

In this chapter, we're going to cover the following main recipes:

- Setting up Eureka Server
- Integrating an application in Eureka Server
- Scaling out the RESTful API
- Setting up Spring Cloud Gateway
- Testing Spring Cloud Gateway
- Setting up Spring Cloud Config
- Protecting Spring Cloud Gateway
- Integrating distributed tracing with Spring Cloud
- Deploying Spring Boot Admin

Technical requirements

This chapter requires some services to be running on your computer, such as OpenZipkin. As usual, the easiest way to run them on your computer is by using Docker. You can get Docker from the Docket product page at `https://www.docker.com/products/docker-desktop/`. I will explain how to deploy each tool in its corresponding recipe.

The *Setting up Spring Cloud Config* recipe requires a git repository. You can create a GitHub account for free (`https://github.com/join`).

You can find the code for all the recipes in this chapter here: `https://github.com/PacktPublishing/Spring-Boot-3.0-Cookbook/tree/main/chapter4`.

Setting up Eureka Server

Eureka Server is a service registry that's used in microservices architectures to register instances that other applications can discover. It's a valuable service that allows services to locate and communicate with each other dynamically. This service registry performs health checks on the registered services' instances, automatically removing the unhealthy or unresponsive ones. When a service needs to communicate with another service, Eureka Server provides the available instances, allowing load balancing.

In this recipe, you will learn how to create an application that implements Eureka Server.

Getting ready

This recipe doesn't have any additional requirements.

How to do it...

In this recipe, we'll create a new Eureka Server that we'll reuse in the rest of the recipes. Let's get started:

1. First, we'll create a new application for Eureka Server. For that, open `https://start.spring.io` and use the same parameters that you did in the *Creating a RESTful API* recipe of *Chapter 1*, except change the following options:

    *   For **Artifact**, type `registry`

    *   For **Dependencies**, select **Eureka Server**

2. Then, in the generated project, in the `resources` folder, create a file named `application.yml` and set the following configuration:

    ```
    server:
        port: 8761
    eureka:
        client:
    ```

```
              registerWithEureka: false
              fetchRegistry: false
```

3. Next, open the `RegistryApplication` class and annotate it with `@EnableEurekaServer`:

    ```
    @EnableEurekaServer
    @SpringBootApplication
    public class RegistryApplication
    ```

4. Now, you can start the application.

5. Let's verify that Eureka Server is running. Open `http://localhost:8761` in your browser:

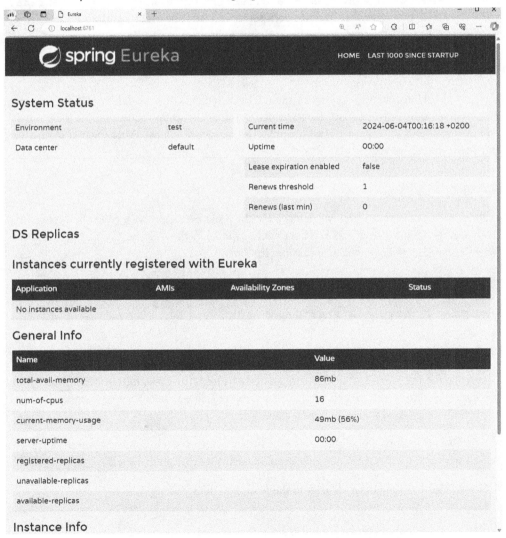

Figure 4.1: Eureka Server

On the Eureka Server page, you can see general information about the server and, most importantly, the applications registered on the server. Now, we don't have any applications registered yet. Once we connect the applications in the following recipes, we will see them under **Instances currently registered with Eureka**.

How it works...

The Eureka Server dependency in a Spring Boot application allows you to set up and run a service registry. When you use the `@EnableEurekaServer` annotation, the Eureka Server autoconfiguration is activated. The Eureka Server application must be configured so that it can stop itself from being registered as a service, which is why the `eureka.client.registerWithEureka` and `eureka.client.fetchRegistry` settings are both set to `false`. The other required Eureka Server configuration is the port. We configured Eureka Server to listen on port `8761`.

Integrating an application in Eureka Server

In this recipe, we'll integrate two applications into Eureka Server, which we deployed in the previous recipe. One application provides football data, which the other application consumes. We'll use Eureka Server to register both applications, at which point the consumer will use Eureka Server to discover the provider application.

Getting ready

In addition to Eureka Server, which we deployed in the previous recipe, we'll reuse the applications we created in the *Defining responses and data model exposed by the API* and *Consuming a RESTful API from another Spring Boot application* recipes in *Chapter 1*.

As a starting point, you can use the applications that I've prepared in this book's repository: `https://github.com/PacktPublishing/Spring-Boot-3.0-Cookbook`. You can find the code in the `chapter4/recipe4-2/start` folder.

How to do it...

We'll integrate the `football` and `albums` applications from the *Consuming a RESTful API from another Spring application using RestClient* recipe in *Chapter 1*, into Eureka Server, which we deployed in the previous recipe. Let's make the required adjustments:

1. First, we will modify the applications so that they connect to the Eureka Server instance. We will start with the `football` application. Make the following changes:

    -   Add the following dependencies to the `pom.xml` file:

        ```
        <dependency>
            <groupId>org.springframework.cloud</groupId>
        ```

```
            <artifactId>spring-cloud-starter-openfeign</artifactId>
    </dependency>
    <dependency>
            <groupId>org.springframework.cloud</groupId>
            <artifactId>spring-cloud-starter-netflix-eureka- client</
    artifactId>
    </dependency>
```

- Ensure the pom.xml file has configured dependency management for Spring Cloud:

```
<dependencyManagement>
    <dependencies>
            <dependency>
                    <groupId>org.springframework.cloud</groupId>
                    <artifactId>spring-cloud-dependencies</
    artifactId>
                    <version>${spring-cloud.version}</version>
                    <type>pom</type>
                    <scope>import</scope>
            </dependency>
    </dependencies>
</dependencyManagement>
```

- Ensure that the spring-cloud.version property is defined in the pom.xml file:

```
<properties>
    <java.version>21</java.version>
    <spring-cloud.version>2022.0.4</spring-cloud.version>
</properties>
```

- In the resources folder, add a file named application.yml with the following content:

```
server:
  port: 0
spring:
  application:
    name: FootballServer
eureka:
  client:
    serviceUrl:
      defaultZone: http://localhost:8761/eureka/
```

2. Start the football application.

3.   At this point, you'll be able to see the application registered in Eureka Server:

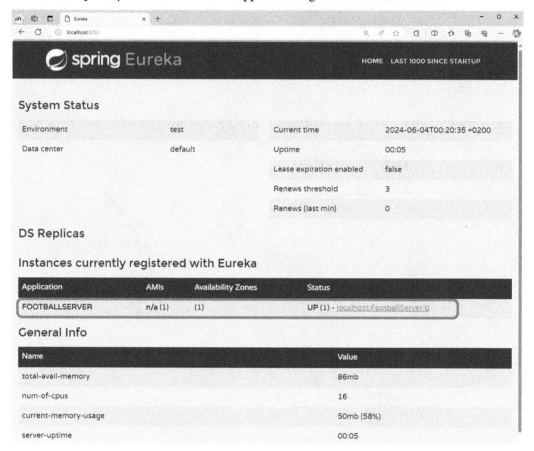

Figure 4.2: The RESTful application registered in Eureka Server

4.   Next, modify the RESTful API `albums` consumer application by making the following changes:

- Add the `org.springframework.cloud:spring-cloud- starter-netflix-eureka-client` and `org.springframework.cloud:spring-cloud-starter-openfeign` dependencies to the `pom.xml` file:

```
<dependency>
      <groupId>org.springframework.cloud</groupId>
      <artifactId>spring-cloud-starter-netflix-eureka-client</artifactId>
</dependency>
<dependency>
```

```
        <groupId>org.springframework.cloud</groupId>
        <artifactId>spring-cloud-starter-openfeign</artifactId>
</dependency>
```

- In the AlbumsApplication. class, add the @EnableDiscoveryClient annotation:

```
@EnableDiscoveryClient
@EnableFeignClients
@SpringBootApplication
public class AlbumsApplication {
```

- In the resources folder, add an application.yml file with the following configuration:

```
spring:
  application:
    name: AlbumsServer
eureka:
  client:
    serviceUrl:
      defaultZone: http://localhost:8761/eureka/
```

- Modify the FootballClient class, changing @FeignClient by setting just the target application name:

```
@FeignClient("FootballServer")
public interface FootballClient {
    @RequestMapping(method = RequestMethod.GET, value = "/
players")
    List<Player> getPlayers();
}
```

Note that we no longer use the remote RESTful API server address, just the application name.

5. Now, you can run the albums application.

6. Finally, you can test the entire deployment. For that, execute the following curl request:

```
curl http://localhost:8080/albums/players
```

The consumer application will discover which instances of the server application are available by asking Eureka Server, after which point it will call the server application and return the result.

How it works...

To set up the client connection to Eureka Server, it is necessary to add the `org.springframework.cloud:spring-cloud-starter-openfeign` and `org.springframework.cloud:spring-cloud-starter-netflix-eureka-client` dependencies and configure the connection to Eureka Server. The configuration on the client side consists of the following:

- `eureka.client.serviceUrl.defaultZone`: This is the address of Eureka Server. In our case, this is `http://localhost:8761/eureka`.

- `spring.appication.name`: This is the name that can be used to discover the service.

OpenFeign and Eureka clients use Eureka Server to discover instances of a service. Remember that in the `@OpenFeignClient` configuration, we used the server application name instead of a server address. The OpenFeign client connects to Eureka Server, requests the instances that have been registered for that service, and returns one.

For clients, this is more straightforward as knowing the address where the server instances will be hosted in advance is unnecessary.

The discovery mechanism is also very convenient for server applications as they don't need to be hosted in a predefined server and port. You probably noticed that the RESTful API server was configured with `server.port=0`, which means it will start in a random port. The server address and port are stored when it's registered in Eureka Server. When the consumer application asks for Eureka Server, it returns information about the registered instance – that is, the server address and port. This feature is helpful as we run our applications locally and we don't need to care about which port we are running each instance on. In previous recipes, we started one application on port `8080` and another on `8081`. In the *Scaling out the RESTful API service* recipe, we will see that it is possible to have more than one instance of a given service.

There's more...

A key feature of Eureka Server is detecting unhealthy or unresponsive application instances and removing them from the registry. This feature requires that registered services use **Actuator**. Spring Actuator provides production-ready features that help you monitor and manage your Spring applications. It's particularly useful for microservices and other distributed systems, where operational visibility and management are critical. You can include the Actuator dependency in your projects by applying the following code:

```
<dependency>
    <groupId>org.springframework.boot</groupId>
    <artifactId>spring-boot-starter-actuator</artifactId>
</dependency>
```

You can find more information about Actuator on the project page: `https://docs.spring.io/spring-boot/docs/current/reference/html/actuator.html`.

Scaling out the RESTful API

Scaling out is a technique that improves the availability and capacity of a system by adding multiple instances for a given service.

In modern application platforms, such as container orchestrators such as Kubernetes or cloud providers hosting platforms such as Azure App Services or AWS Elastic Beanstalk, the systems may scale out and scale in automatically. For instance, in Kubernetes, you can configure an autoscale rule that increases the number of instances of your service when the average CPU has been over 70% for the last 5 minutes. You can also configure it in another way – when the usage of your application is low, you can scale in your application. This means you can decrease the number of instances of the application.

Scaling out an application shouldn't necessarily be automated; you can scale it manually, as we'll do in this recipe.

Scaling out involves distributing incoming requests across multiple instances of a service. In this recipe, we will learn how to use Eureka Server capabilities to register and discover instances to distribute the requests across available service instances.

Getting ready

In this recipe, we will use the services we utilized in the previous recipe:

- **Eureka Server**: This service will act as a service registry and provide service discovery
- **RESTful API**: This will provide a service to be consumed by the client application
- **Client application**: This will consume the RESTful API

If you haven't completed the previous recipe, you can find the completed exercise in this book's GitHub repository at `https://github.com/PacktPublishing/Spring-Boot-3.0-Cookbook`.

You can find the code to start this recipe in the `chapter4/recipe4-3/start` folder.

How to do it...

We will modify the RESTful API so that it returns the service instance's information. That way, we can validate that the requests are balanced among available instances. Then, we will execute more than one instance of the RESTful API. Let's get started:

1. In the RESTful API project, modify the `application.yml` file in the `resources` folder by adding the following property at the beginning of the file:

    ```
    instance:
        instance-id: ${spring.application.name}:${random.int}
    ```

The file should look like this:

```
football:
  instanceId: ${random.uuid}
server:
  port: 0
spring:
  application:
    name: FootballServer
eureka:
  client:
    serviceUrl:
      defaultZone: http://localhost:8761/eureka/
  instance:
    instance-id: ${spring.application.name}:${random.int}
```

2. Create a new controller named `ServiceInformationController` and write the following code:

```
@RequestMapping("/serviceinfo")
@RestController
public class ServiceInformationController {
    @Value("${football.instanceId}")
    private String instanceId;
    @GetMapping
    public String getInstanceId() {
        return instanceId;
    }
}
```

3. Execute three instances of the RESTful API. Instead of using `mvnw spring-boot:run`, we will build the JAR file and execute it using the Java runtime. To do this, follow these steps:

    I.   In the root folder of the project, build the application using the following command:

    ```
    ./mvnw package
    ```

    II. Then, open three Terminals and execute the following command in all of them:

    ```
    java -jar ./target/football-0.0.1-SNAPSHOT.jar
    ```

III. Open Eureka Server at `localhost:8761`. You'll see three instances of the RESTful API service:

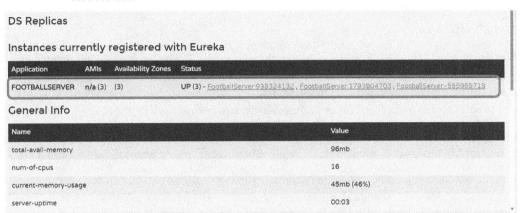

Figure 4.3: Eureka Server with three instances of FootballServer running

4. In the client application project, make the following changes:

- In the `FootballClient` class, add the following method:

```
@RequestMapping(method = RequestMethod.GET, value="/
serviceinfo")
String getServiceInfo();
```

- In the `AlbumsController` controller, add the following method:

```
@GetMapping("/serviceinfo")
public String getServiceInfo(){
    return footballClient.getServiceInfo();
}
```

5. Now, start the client application and test the application multiple times. You can do that by executing the following `curl` request several times:

```
curl http://localhost:8080/albums/serviceinfo
```

You will see that the results change when you execute the previous command multiple times:

```
bili@DESKTOP-GSIKJL4:~/repos/albums$ curl http://localhost:8080/albums/serviceinfo
227d49d6-33a5-4830-967a-2dffcc9302cabili@DESKTOP-GSIKJL4:~/repos/albums$ curl http://localhost:8080/albums/serviceinfo
4f353ad1-3df0-40ec-b89b-9f9e5003e772bili@DESKTOP-GSIKJL4:~/repos/albums$ curl http://localhost:8080/albums/serviceinfo
c149b910-197d-4efe-bcce-f91cf111cd18bili@DESKTOP-GSIKJL4:~/repos/albums$ curl http://localhost:8080/albums/serviceinfo
227d49d6-33a5-4830-967a-2dffcc9302cabili@DESKTOP-GSIKJL4:~/repos/albums$ curl http://localhost:8080/albums/serviceinfo
c149b910-197d-4efe-bcce-f91cf111cd18bili@DESKTOP-GSIKJL4:~/repos/albums$ curl http://localhost:8080/albums/serviceinfo
4f353ad1-3df0-40ec-b89b-9f9e5003e772bili@DESKTOP-GSIKJL4:~/repos/albums$ curl http://localhost:8080/albums/serviceinfo
227d49d6-33a5-4830-967a-2dffcc9302cabili@DESKTOP-GSIKJL4:~/repos/albums$
```

Figure 4.4: Results of executing the RESTful API from the client application

The client application distributes requests across Eureka Server's registered service instances, resulting in different outcomes.

How it works...

When the Eureka client starts, it registers itself in Eureka Server. The registration details include the service name and network location. After the registration process, the client sends heartbeats to inform the server that it is still alive. In this exercise, we initiated three instances of the RESTful API server with the same service name; each of the instances had a separate network location.

The Feign client in the consumer application uses Eureka Server to discover the available instances of the RESTful API server application. In that way, it can balance the request across the service instances.

Just for demonstration purposes, we added a configuration setting, `football.InstanceId`, with a unique random value to distinguish the service instance. To retrieve that configuration, we used the `@Value` annotation. Spring Boot injected the value when the application was started.

Setting up Spring Cloud Gateway

When creating complex applications with different services, we don't want to expose all those services to consumer applications so that we can avoid unnecessary complexity exposure. To address this scenario, we can use **Spring Cloud Gateway**. Spring Cloud Gateway can be deployed in such a way that it's the only component that's accessible to consumer applications, while the rest of the services will be accessed either internally or just from Spring Cloud Gateway. This is illustrated in *Figure 4.5*:

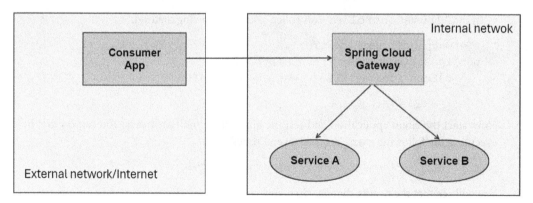

Figure 4.5: A typical Spring Cloud Gateway deployment

> **A note on deployment**
>
> Depending on the complexity and requirements of the solution, I recommend using additional networking protections, such as Layer 7 load balancers, **web application firewalls (WAFs)**, or other protection mechanisms. For learning purposes, I will not describe them in this book and focus on Spring and Spring Cloud mechanisms instead.

In addition to the role of an API gateway offering a unique entry point for the application, Spring Cloud Gateway has interesting benefits:

- **Load balancing**: It can balance requests across the available service instances.
- **Dynamic routing**: Spring Cloud Gateway can be integrated with a service registry, such as Eureka Server, and dynamically route requests.
- **Security**: It can apply authentication and authorization using authentication providers, such as Spring Security and OAuth2, and propagate to downstream services. You can do this in a single place if you need to configure CORS for your consumer application.
- **SSL termination**: You can configure Spring Cloud Gateway to terminal SSL/TLS connections and pass unencrypted traffic to the services. With this feature, you can offload the SSL/TLS decryption from the services.
- **Rate limiting**: You can implement rate limiting to prevent your services from being abused.
- **Request/response transformation**: You can use Spring Cloud Gateway to transform requests and responses – for instance, by adding requests or response headers. You can also convert payload formats, such as XML, into JSON. These transformations can be applied at the gateway level; hence, it is not necessary to modify your downstream services.
- **Circuit breaking**: You can use Spring Cloud Gateway to implement circuit breakers to handle failures gracefully. For instance, you can prevent requests from being sent to an unhealthy service.

Some added benefits are *request filtering, global exception handling, logging and monitoring,* and *path Rewriting.*

I recommend visiting the project page at `https://spring.io/projects/spring-cloud-gateway` for more details.

In this recipe, we will deploy an instance of Spring Cloud Gateway and integrate it with Eureka Server, which we deployed in previous recipes, to route requests to registered services.

Getting ready

In this recipe, we will use the projects we implemented in previous recipes:

- **Eureka Server**: This service will act as a service registry and provide service discovery.

- **RESTful API**: This will provide a service to be consumed by the client application – that is, the `football` application.

- **Consumer API**: This application will consume the RESTful API service. We will modify the application so that it provides an additional endpoint in this recipe. This is the `album` application.

If you haven't completed the previous recipe, you can find the completed exercise in this book's GitHub repository at `https://github.com/PacktPublishing/Spring-Boot-3.0-Cookbook`. The code to start this recipe can be found in the `chapter4/recipe4-4/start` folder.

How to do it...

Let's deploy a Spring Cloud Gateway. We will configure Gateway so that it exposes some functionality of the RESTful API:

1. Open `https://start.spring.io` and use the same parameters that you did in the *Creating a RESTful API* recipe, except change the following options:

    -   For **Artifact**, type `gateway`

    -   For **Dependencies**, select **Gateway** and **Eureka Discovery Client**

2. In the project you've downloaded, create a file in the `resources` folder named `application.yml` with the following content:

    ```yaml
    spring:
      application:
        name: GatewayServer
      cloud:
        gateway:
          routes:
            - id: players
              uri: lb://footballserver
              predicates:
                - Path=/api/players/**
              filters:
                - StripPrefix=1
    eureka:
      client:
    ```

```
serviceUrl:
    defaultZone: http://localhost:8761/eureka/
```

3. Now, you can run the gateway application. One important thing to note is that the other application should be running as well.

4. Test the gateway by executing the following request:

```
curl http://localhost:8080/api/players
```

You should see the RESTful API's results.

5. Now, let's add a new method in the other RESTful API applications in `Albums` and then add it as a new route in Spring Cloud Gateway. So, open the `AlbumsController` controller and add the following method:

```
@GetMapping
public List<String> getAlbums(){
    return List.of("Album 1", "Album 2", "Album 3");
}
```

6. In the same project, open the `application.yml` file and add the following property:

```
server:
    port: 0
```

Now, add a new route in the Spring Cloud Gateway configuration. For that, open the `application.yml` file of the Spring Cloud Gateway project and add the following highlighted text. I've added the entire configuration file for clarity:

```
spring:
  application:
    name: GatewayServer
  cloud:
    gateway:
      routes:
        - id: players
          uri: lb://footballserver
          predicates:
            - Path=/api/players/**
          filters:
            - StripPrefix=1
        - id: albums
          uri: lb://albumsserver
          predicates:
            - Path=/api/albums/**
          filters:
            - StripPrefix=1
```

```
eureka:
  client:
    serviceUrl:
      defaultZone: http://localhost:8761/eureka/
```

7. Restart Spring Cloud Gateway and test the new route by executing the following `curl` request:

    ```
    curl http://localhost:8080/api/albums
    ```

    Now, you should see the response of the second RESTful API.

How it works...

In this recipe, we connected Spring Cloud Gateway to Eureka Server. For that, we only needed to include the Eureka discovery client and its configuration – that is, `eureka.client.serviceUrl.defaultZone` property in the `application.yml` file.

Once connected to Eureka Server, we configured a couple of routings. A routing definition specifies a combination of criteria and actions to be taken when a request matches the criteria.

We established the criteria for route definition by employing predicates. Specifically, we configured two routes: one using the `/api/players/**` path pattern and the other using `/api/albums/**`. This configuration dictates that the first route will match requests starting with `/api/player`, while the second route will match requests commencing with `/api/albums`. For example, a request such as `http://localhost:8080/api/player` would match the first route. Beyond the request path, you can utilize other request properties, such as headers, query parameters, or the request host.

Since the target services expect the requests as `/players` in one case and `/albums` in the other, without `/api` in both cases, removing this part of the path is necessary. We configured this using the `StripPrefix=1` filter, which removed the first part of the path.

Finally, those routes needed to hit a target service, so we configured this using the `uri` property. We could have used the DNS host and port, something like `http://server:8081`, but instead, we used `lb://servicename`. Using this approach, we configured Spring Cloud Gateway to discover the target service using Eureka and leverage client-side load balancing. We deployed all our services locally, and the only way to distinguish each instance is by dynamically assigning each service a port.

> **Note**
> To assign a port dynamically, we set the `server.port=0` property.

If the hosting environment provides an alternative balancing method, it is acceptable to use it. For instance, in a Kubernetes environment, you can create a deployment for your service with multiple running instances. By doing this, your service can be discovered through Kubernetes DNS, and the underlying infrastructure will balance requests.

See also

I recommend reading Spring Cloud Gateway documentation, which you can find here: `https://spring.io/projects/spring-cloud-gateway`. Familiarize yourself with routing capabilities and understand how to configure your routes using all the properties that are available in the requests.

Circuit Breaker is also an interesting design pattern that can be very useful for handling failures gracefully. If you are unfamiliar with this pattern, I recommend looking at this *Azure Cloud Design Patterns* article: `https://learn.microsoft.com/azure/architecture/patterns/circuit-breaker`. The good news is that this pattern is relatively easy to implement using Spring Cloud Gateway – see `https://spring.io/guides/gs/gateway/` for more details.

Testing Spring Cloud Gateway

As the Spring Cloud Gateway rules are processed at runtime, they can sometimes be difficult to test. In addition to the rules themselves, the target applications must be up and running.

In this recipe, we'll learn how to test Spring Cloud Gateway using the *Spring Cloud Contract Stub Runner* starter, which emulates the target services using the Wiremock library.

Getting ready

In this recipe, we'll create tests for the Spring Cloud Gateway project we set up in the previous recipe. I've prepared a working version of Spring Cloud Gateway in case you haven't set it up yet. You can find it in this book's GitHub repository at `https://github.com/PacktPublishing/Spring-Boot-3.0-Cookbook`. The code to start this recipe can be found in the `chapter4/recipe4-5/start` folder. I've added all the projects that were used in the previous recipe – that is, the `football`, the `albums`, and the `gateway` projects – but we'll only be using `gateway` here.

How to do it...

In this recipe, we'll adjust the gateway project to allow for test execution. Let's get started:

1. First, we'll add the *Spring Cloud Contract Stub Runner* starter. For that, add the following dependency in the `gateway` project's `pom.xml` file:

    ```
    <dependency>
        <groupId>org.springframework.cloud</groupId>
        <artifactId>spring-cloud-starter-contract-stub-runner</artifactId>
        <scope>test</scope>
    </dependency>
    ```

    Note that this dependency is used for testing purposes only.

2. Next, modify the `application.yml` configuration to parameterize the destination URIs. Replace the addresses in `spring.cloud.gateway.routes.uri` so that they use a configuration parameter:

```yaml
spring:
  application:
    name: GatewayServer
  cloud:
    gateway:
      routes:
        - id: players
          uri: ${PLAYERS_URI:lb://footballserver}
          predicates:
            - Path=/api/players/**
          filters:
            - StripPrefix=1
        - id: albums
          uri: ${ALBUMS_URI:lb://albumsserver}
          predicates:
            - Path=/api/albums/**
          filters:
            - StripPrefix=1
eureka:
  client:
    serviceUrl:
      defaultZone: http://localhost:8761/eureka/
```

3. Before creating our first test, we need to set up the test class. Let's create a new class named `RoutesTests` in the `test` folder. To set it up, you must do the following:

-   Annotate the class with `@AutoConfigureWireMock(port = 0)`

-   Annotate the class with `@SpringBootTest` using the `properties` field to pass the destination URIs

-   Add the `WebTestClient` field that the Spring Boot tests context will inject

The skeleton of this class should look like this:

```java
@AutoConfigureWireMock(port = 0)
@SpringBootTest(webEnvironment = SpringBootTest.WebEnvironment.
RANDOM_PORT, properties = {
        "PLAYERS_URI=http://localhost:${wiremock.server.port}",
        "ALBUMS_URI=http://localhost:${wiremock.server.port}",
})
public class RoutesTester {
```

```
    @Autowired
    private WebTestClient webClient;
}
```

4. Now, we can create our first test. We'll add a new method annotated with @Test to check the players route:

I. Name the method playersRouteTest:

```
@Test
public void playersRouteTest() throws Exception
```

II. First, arrange the response of the target server when calling the /players path. We'll use the Wiremock library:

```
stubFor(get(urlEqualTo("/players"))
        .willReturn(aResponse()
                .withHeader("Content-Type", "application/json")
                .withBody("""
                    [
                        {
                            "id": "325636",
                            "jerseyNumber": 11,
                            "name": "Alexia PUTELLAS",
                            "position": "Midfielder",
                            "dateOfBirth": "1994-02-04"
                        },
                        {
                            "id": "396930",
                            "jerseyNumber": 2,
                            "name": "Ona BATLLE",
                            "position": "Defender",
                            "dateOfBirth": "1999-06-10"
                        }
                    ]""")));
```

III. Now, we can call Spring Cloud Gateway by using WebTestClient and assert that it's working as expected:

```
webClient.get().uri("/api/players").exchange()
        .expectStatus().isOk()
        .expectBody()
        .jsonPath("$[0].name").isEqualTo("Alexia PUTELLAS")
        .jsonPath("$[1].name").isEqualTo("Ona BATLLE");
```

5.   Now, you can test the `albums` route using the same approach. This book's GitHub repository contains more tests for Spring Cloud Gateway: `https://github.com/PacktPublishing/Spring-Boot-3.0-Cookbook`.

How it works...

When configuring the Spring Cloud Gateway project, two dependencies need to be considered: Eureka Server and the target RESTful API. However, the main purpose is to verify the Gateway routes during testing. To achieve this, we removed the dependency on Eureka Server and allowed the target RESTful API URI to be configured. By using the `${key:default}` notation in *Step 2*, we created a fallback mechanism that uses the configured value for the load balancer address. If no value is provided, then it defaults to the original URI. This notation specifies that if the key is provided, then it uses that key; otherwise, it uses the default value specified after the colon symbol.

Using the configuration mechanism described previously and the Wiremock provided by the *Spring Cloud Contract Stub Runner* starter, we configured the address of the remote RESTful APIs, considering that the Wiremock server is running on localhost and the port is provided by the Wiremock server. In the `@AutoConfigureWireMock` annotation, we used port 0 to ensure the port is assigned randomly. Then, using `${wiremock.server.port}`, we retrieved the assigned port.

The rest of the test follows the same mocking mechanism that we explained in the *Mocking a RESTful API* recipe in *Chapter 1*. Note that the mocked RESTful API responds to `/players`, while the test requests `/api/players`. In this test, we want to validate that the Spring Cloud Gateway configuration is correct, so when making a request to `/api/players`, it redirects the call to the target API on the `/players` path. So long as the test is implemented correctly and Spring Cloud Gateway is configured properly, the test should pass without any issues.

Setting up Spring Cloud Config

Spring Cloud Config enables centralized configuration management for applications, allowing you to store configuration properties in a central repository and distribute them to connected services.

It provides the following features, among others:

- It allows version control configurations – for instance, using git as a backend to store the configuration. With this feature, you can track changes and audit configurations, and facilitate performing rollbacks to previous versions when needed.

- It enables dynamic configuration updates with no need to restart services.

- It externalizes the configuration; hence, it is possible to make configuration changes without modifying or redeploying the services.

In this recipe, we will deploy a configuration server and connect our existing RESTful APIs to the configuration service.

Getting ready

For this recipe, you will need a Git repository. I recommend using GitHub as this recipe has been tested and validated with this service, but I don't foresee any issue if you use another git provider. If you want to use GitHub and don't have an account yet, visit `https://github.com`. You will also need a git client.

I will reuse the RESTful APIs that we configured in the previous recipe. These are the services we must configure:

- `football` (RESTful API)
- `albums` (RESTful API)
- `gateway`

If you haven't completed the previous recipe yet, you can use the completed recipe in this book's GitHub repository: `https://github.com/PacktPublishing/Spring-Boot-3.0-Cookbook`.

The code to start this recipe can be found in the `chapter4/recipe4-6/start` folder.

How to do it...

In this recipe, we'll create a new service using Spring Initializr to host Spring Cloud Config. Next, we'll configure the service to use a GitHub repository as a backend. Finally, we'll connect existing services to the Config server. Let's get started:

1. Open `https://start.spring.io` and use the same parameters that you did in the *Creating a RESTful API* recipe in *Chapter 1*, except changing the following options:

   - For **Artifact**, type `config`
   - For **Dependencies**, select **Config Server**

2. Create a GitHub repository in your GitHub account. Since we won't be managing anything secretive, the repository can be public. Name it `spring3-recipes-config`.

3. Clone the repository on your computer. To do that, open a Terminal and execute the following command, replacing `felipmiguel` with your GitHub account's name:

   ```
   git clone https://github.com/felipmiguel/spring3-recipes-config
   ```

   This will create `spring3-recipes-config` as the root folder for that repository.

   In the following steps, we will create files in that folder that will later be pushed to GitHub's central repository.

4. In the root folder of the configuration repository, create the following files:

    - `application.yml`, with the following content:

    ```yaml
    server:
        port: 0
    eureka:
      client:
        serviceUrl:
          defaultZone: http://localhost:8761/eureka/
        instance:
          instance-id: ${spring.application.name}:${random.int}
    ```

    - `gatewayserver.yml`, with the following content:

    ```yaml
    server:
      port: 8080
    spring:
      cloud:
        gateway:
          routes:
            - id: players
              uri: ${PLAYERS_URI:lb://footballserver}
              predicates:
                - Path=/api/players/**
              filters:
                - StripPrefix=1
            - id: albums
              uri: ${ALBUMS_URI:lb://albumsserver}
              predicates:
                - Path=/api/albums/**
              filters:
                - StripPrefix=1
    ```

5. Next, push the files to `github.com`. To do this, execute the following commands in your Terminal in the repository root folder:

    ```
    git commit -m "Initial configuration" .
    git push
    ```

6. Configure your repository as the backend of the Config service. For this, go to the Config Service project and add a file named `application.yml` in the `resources` folder with the following content (make sure you replace [your account] with your GitHub account's name):

```
server.port: 8888
spring:
  cloud:
    config:
      server:
        git:
          uri: https://github.com/[your account]/spring3-
recipes-config
```

7. Open the application's `ConfigApplication` class and add the `@EnableConfigServer` annotation. It should look like this:

```
@EnableConfigServer
@SpringBootApplication
public class ConfigApplication
```

8. Now, you can start the Config server.

9. Next, modify the projects so that you can connect to the Config server. To do so, follow these steps:

   - Add dependencies to the `pom.xml` file for all the applications we want to connect to the Config server. These applications are `football`, `album`, `registry`, and `gateway`:

   ```
   <dependency>
         <groupId>org.springframework.cloud</groupId>
         <artifactId>spring-cloud-starter-config</artifactId>
   </dependency>
   ```

   - Configure the `application.yml` file for all the applications we want to connect to the Config server. All of them will contain the Config server configuration and the respective application's name. For instance, the `album` service will look like this:

   ```
   spring:
     config:
       import: optional:configserver:http://localhost:8888
     application:
       name: AlbumsServer
   ```

- For the `football` service (the RESTful API service), set the following content:

```
football:
  instanceId: ${random.uuid}
spring:
  config:
    import: optional:configserver:http://localhost:8888
  application:
    name: FootballServer
```

- For the `gateway` service, set the following content:

```
spring:
  config:
    import: optional:configserver:http://localhost:8888
  application:
    name: gatewayserver
```

10. Now, it's time to verify that everything is working. Let's start all services.

11. Test the services by executing a request to Spring Cloud Gateway:

```
curl http://localhost:8080/api/players
```

12. Validate that it returns a JSON file containing a list of players.

How it works...

Spring Boot provides an extensible mechanism to load the configuration from external sources using the `spring.config.import` setting. Adding the `org.springframework.cloud:spring-cloud-starter-config` dependency registers an extension that can retrieve the configuration from a config server.

To set up the configuration server, the only requirement is adding the `org.springframework.cloud:spring-cloud-config-server` dependency and enabling the configuration server using the `@EnableConfigServer` annotation. Enabling the configuration server exposes an endpoint that allows consumer applications to query for their configuration. The configuration endpoint exposes the following paths:

```
/{application}/{profile}[/{label}]
/{application}-{profile}.yml
/{label}/{application}-{profile}.yml
/{application}-{profile}.properties
/{label}/{application}-{profile}.properties
```

Let's take a look at each path fragment:

- `application` is the application name that's configured by the `spring.application.name` property.

- `profile` is the currently active profile. By default, the profile's name is `default`.

- `label` refers to a git branch; if not specified, it applies to the default branch.

Our applications provide the following queries for the Config server:

- `football`: As it contains the `spring.application.name=FootballServer` property, it requests `http://localhost:8888/FootballServer-default.yml`

- `albums`: Its application name is `AlbumsServer`, so it requests `http://localhost:8888/AlbumsServer-default.yml`

- `gateway`: Its application name is `GatewayServer`, so it requests `http://localhost:8888/GatewayServer-default.yml`

You can see the results by executing a request. For instance, for `GatewayServer`, you can run the following command:

```
curl http://localhost:8888/GatewayServer-default.yml
```

The result should look like this:

```
server:
  port: 8080
eureka:
  client:
    serviceUrl:
      defaultZone: http://localhost:8761/eureka/
spring:
  cloud:
    gateway:
      routes:
      - id: players
        uri: ${PLAYERS_URI:lb://footballserver}
        predicates:
        - Path=/api/players/**
        filters:
        - StripPrefix=1
      - id: albums
        uri: ${ALBUMS_URI:lb://albumsserver}
```

```
        predicates:
        - Path=/api/albums/**
        filters:
        - StripPrefix=1
```

Let's analyze what the Config server did. The Config server resolves the configuration by merging the configurations it found in the git repository:

- The base configuration starts with the `application.yml` file.

- It merges the base configuration with a more specific configuration for the requested application. The more specific configuration is defined using a `[application name].yml` file, where `[application name]` is defined in the `spring.application.name` property. In our scenario, we haven't defined specific configuration files for the `football` and `albums` applications, but we did define the `gatewayserver.yml` file for the `gateway` service. By doing this, `gateway` will merge the content of `application.yml` and `gatewayserver.yml`.

- If settings are defined in multiple files, the most specific one is used. In this case, the settings defined by `gatewayserver.yml` will take precedence over the settings defined in `application.yml`. You can see this behavior for the `server.port` setting, which is specified in both files and takes the most specific one.

There's more...

In production environments, you probably want to protect your applications' configurations. For that reason, you must use a private git repository, your configuration service will require authentication, and your secrets, such as connection strings, will be encrypted. You can do all that using Spring Cloud Config. I recommend visiting the project page at `https://spring.io/projects/spring-cloud-gateway` for details on the configuration.

Another exciting feature related to configuration is the possibility to dynamically refresh the configuration without restarting the application. You can achieve this by using Spring Actuator. We will revisit this topic in later chapters.

We just used non-sensitive information in this recipe, but applications usually manage configurations we don't want to disclose, such as database connection strings or credentials to access other systems.

The first measure we should apply is removing public access to the configuration repository. We can use private repositories and configure the git credentials in the Config server like so:

```
spring:
  cloud:
    config:
      server:
        git:
          uri: https://github.com/PacktPublishing/Spring-Boot-3.0-
Cookbook-Config
          username: theuser
          password: strongpassword
```

To avoid storing sensitive information in a git repository, Spring Cloud Config has an extension to integrate with Vault services, such as Hashicorp Vault and Azure Key Vault. The configuration file that's stored in the git repository contains references to secrets stored in the Vault service. The applications resolve the configuration, retrieving the referenced secrets from the Vault service.

See also

See the Spring Cloud Config quickstart guide at `https://docs.spring.io/spring-cloud-config/docs/current/reference/html/` for more advanced scenarios.

Integrating distributed tracing with Spring Cloud

As the number of services composing the `football` application suite grew, you deployed the following Spring Cloud components: Spring Cloud Gateway, Eureka Server (a registry and discovery service), and Spring Cloud Configuration. You want to configure distributed tracing to monitor the transactions across microservices.

In this recipe, you will integrate distributed tracing with Actuator and OpenZipkin into a system composed of different application microservices and Spring Cloud components.

Getting ready

You will monitor distributed transactions using OpenZipkin. As explained in the *Implementing distributed tracing* recipe in *Chapter 3*, you can deploy an OpenZipkin server on your computer using Docker. For that, you can run the following command in your Terminal:

```
docker run -d -p 9411:9411 openzipkin/zipkin
```

You will reuse the outcome of the *Setting up Spring Cloud Config* recipe. I've prepared a working version in case you haven't completed that recipe yet. You can find it in this book's GitHub repository at `https://github.com/PacktPublishing/Spring-Boot-3.0-Cookbook/`, in the `chapter4/recipe4-7/start` folder. It includes the following projects:

- `config`: The Spring Cloud Config service.

- `registry`: The Spring Cloud registry and discovery service.

- `gateway`: Spring Cloud Gateway. It exposes the `football` and `albums` services.

- `football`: The `football` service, which provides information about teams and players.

- `albums`: The `albums` service, which manages sticker albums. It uses the `football` service.

How to do it...

Let's configure our Spring Cloud solution so that we can integrate distributed tracing with OpenZipkin.

1. You must add a dependency to Actuator, the Micrometer bridge to OpenTelemetry, and the exporter from OpenTelemetry to OpenZipkin to all projects. For that, add the following dependencies to all `pom.xml` project files, - that is, the `config`, `registry`, `gateway`, `football`, and `albums` projects:

```
<dependency>
    <groupId>org.springframework.boot</groupId>
    <artifactId>spring-boot-starter-actuator</artifactId>
</dependency>
<dependency>
    <groupId>io.micrometer</groupId>
    <artifactId>micrometer-tracing-bridge-otel</artifactId>
</dependency>
<dependency>
    <groupId>io.opentelemetry</groupId>
    <artifactId>opentelemetry-exporter-zipkin</artifactId>
</dependency>
```

2. The `albums` project also makes some calls to the `football` project using the `OpenFeign` client. For that reason, you should also add the following dependencies to that project:

```
<dependency>
    <groupId>io.micrometer</groupId>
    <artifactId>micrometer-tracing</artifactId>
</dependency>
<dependency>
    <groupId>io.github.openfeign</groupId>
```

```
        <artifactId>feign-micrometer</artifactId>
    </dependency>
```

3. Now, let's change the configuration to enable 100% sampling. Since we're using a central config server, we can change the configuration in the repository that contains the configuration for all applications. In my case, that repository is hosted at `https://github.com/felipmiguel/` `spring3-recipes-config`. As the *Setting up Spring Cloud Config* recipe explains, you should replace `felipmiguel` with your GitHub account. In my repository, I added the following configuration to the `application.yml` file:

```
management:
    tracing:
        sampling:
            probability: 1.0
```

You can create a branch for this feature in the configuration repository. Once you've done this, you will need to modify the configuration in the client applications by adding the following setting in the client `application.yml` file:

```
spring
  cloud:
    config:
      label: <your branch name>
```

Then, you should replace `<your branch name>` with the branch name you created in GitHub.

4. You can now run the applications. You should start the `config` service first, then `registry`, at which point you can start all the rest in no specific order.

5. Let's test the solution. You can run the following requests for testing:

```
curl http://localhost:8080/api/players
```

This request, which is initially handled by the gateway, will be served by the `football` service:

```
curl http://localhost:8080/api/albums
```

Again, this request is initially handled by `gateway` but is served by the `albums` service:

```
curl http://localhost:8080/api/albums/players
```

In this case, the request is initially handled by `gateway` and served by the `albums` service, which simultaneously calls the `football` service.

6. Finally, you can see the traces in OpenZipkin. For that, open `http://localhost:9411` in your browser. Go to **Find a trace** to view the traces. You will see some traces that were initiated in the gateway. These are the ones you executed in *Step 5*:

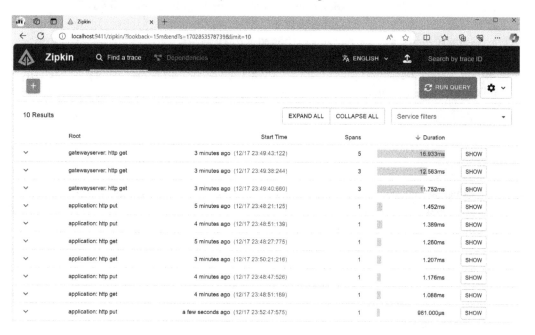

Figure 4.6: Distributed traces for Spring Cloud

The other traces are from the applications synchronizing with Eureka Server.

If you open the traces for `gatewayserver` with five spans – that is, the one corresponding to `/api/albums/players` – you will see that the `gateway` server called the `albums` server, which called the `football` server:

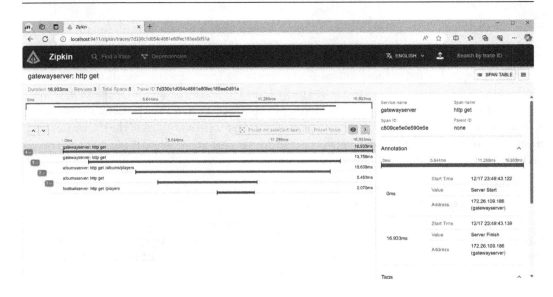

Figure 4.7: Distributed tracing starting in the gateway server, which calls
the albums service, which, in turn, calls the football service

If you open the **Dependencies** section, you will see the dependencies between the microservices:

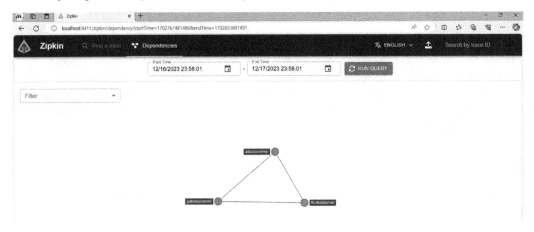

Figure 4.8: The dependencies between Spring Cloud microservices

This view is interesting in complex scenarios with different microservices calling each other
when you need to understand the relationship between them.

How it works...

As explained in the *Implementing distributed tracing* recipe in *Chapter 3*, just by adding Actuator and Micrometer dependencies, the applications send traces to the OpenZipkin server using the default configuration. The default configuration is `http://localhost:9411` for the OpenZipkin server and 10% of sampling. Sampling means that only a percentage of the traces is processed, so only 10% is processed by default. For demonstration purposes, we wanted to send 100% of the traces; for that reason, we took advantage of having a centralized configuration and only changed the `application.yml` file in the configuration repository.

The `albums` application uses an `OpenFeign` client, which, by default, does not propagate the distributed traces as `WebClient.Builder` and `RestTemplateBuilder` do. Therefore, we needed to add two additional dependencies to `io.micrometer:micrometer-tracing` and `io.github.openfeign:feign-micrometer`. On the other hand, Spring Cloud Gateway uses `WebClient.Builder` to make requests to the downstream services. For that reason, the traces are created and propagated correctly from Spring Cloud Gateway with no additional configurations required.

Deploying Spring Boot Admin

After deploying several microservices, you will appreciate having a single dashboard to monitor and manage all of them in one place. Spring Boot Admin is an open source community project that provides a web interface where you can manage and monitor Spring Boot applications.

Getting ready

You will reuse the applications from the *Integrating distributed tracing with Spring Cloud* recipe. I've prepared a working version in case you haven't completed that recipe yet. You can find it in this book's GitHub repository at `https://github.com/PacktPublishing/Spring-Boot-3.0-Cookbook/`, in the `chapter4/recipe4-8/start` folder.

How to do it...

We need to deploy a Spring Boot Admin server and ensure it connects to the discovery service to monitor and manage all applications. Follow these steps:

1. First, create a new application for Spring Boot Admin using the *Spring Initializr* tool. Open `https://start.spring.io` and use the same parameters that you did in the *Creating a RESTful API* recipe of *Chapter 1*, except change the following options:

    *   For **Artifact**, type `fooballadmin`

    *   For **Dependencies**, select **Spring Web**, **Codecentric's Spring Boot Admin (Server)**, **Config Client**, and **Eureka Discovery Client**

2. Next, you must configure Spring Boot Admin. For that, add an `application.yml` file to the `resources` folder with the following content:

```yaml
spring:
  application:
    name: admin-server
  config:
    import: optional:configserver:http://localhost:8888
  cloud:
    config:
      label: distributed-tracing
```

I'm using `spring.cloud.config.label` in this configuration. As I don't want to mix the configuration from different recipes, I've created a new branch for the recipes in this chapter, the name of which is `distributed-tracing`. However, if you made all configurations in the same GitHub repository and the same branch, this setting is not necessary.

3. There is a required additional configuration, but it should be done in the central repository this time since we are using the Spring Cloud Config service. In my case, the configuration is saved at `https://github.com/felipmiguel/spring3-recipes-config`; you should replace `felipmiguel` with your GitHub account, as explained in the *Setting up Spring Cloud Config* recipe. As mentioned in the previous step, I prepared the changes in the `distributed-tracing` branch:

    I.   First, expose Spring Boot Admin through Spring Cloud Gateway. To do so, create a new route in the `gatewayserver.yml` file, as follows:

```yaml
spring:
  cloud:
    gateway:
      routes:
        - id: players
          uri: lb://footballserver
          predicates:
            - Path=/api/players/**
          filters:
            - StripPrefix=1
        - id: albums
          uri: lb://albumsserver
          predicates:
            - Path=/api/albums/**
          filters:
            - StripPrefix=1
```

```
        - id: admin
          uri: lb://admin-server
          predicates:
            - Path=/admin/**
          filters:
            - StripPrefix=1
```

Note that the rest of the routes were already in the configuration.

II.   Next, configure Spring Boot Admin behind Spring Cloud Gateway. To do so, create a file named admin-server.yml in your GitHub repository with the following content:

```
spring:
  boot:
    admin:
      ui:
        public-url: http://localhost:8080/admin
```

III.   Finally, enable some actuator endpoints for all microservices. Since you are using a centralized configuration, you only need to add the following to the application.yml file in your GitHub repository:

```
management:
    endpoints:
        web:
            exposure:
                include:
health,env,metrics,beans,loggers,prometheus
        tracing:
            sampling:
                probability: 1.0
```

4.   The last step before we run the application is configuring the Spring Boot Admin application to enable Admin Server and the Spring Cloud Discovery client. For that, open the FootballAdminApplication class and add the following annotations:

```
@SpringBootApplication
@EnableAdminServer
@EnableDiscoveryClient
public class FootballadminApplication
```

5. Now, you can run the Spring Boot Admin application. Remember that you will need to run the rest of the applications that were reused from the *Integrating distributed tracing with Spring Cloud* recipe and that the `config` and `registry` services should start before the other services. As the Spring Boot Admin service is exposed through Spring Cloud Gateway, you can open `http://locahost:8080/admin` to access Spring Boot Admin:

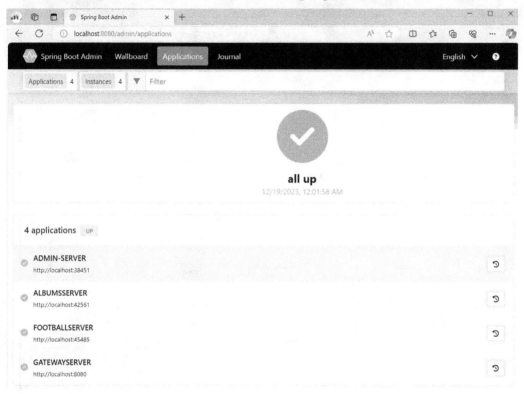

Figure 4.9: The initial Spring Boot Admin page. It defaults to the Applications view

When you access Spring Boot Admin, it redirects you to the **Applications** view. It retrieves the list from Eureka Server. On the application, if you click on the green check on the left-hand side, you will be redirected to the application details page:

Figure 4.10: Application details in Spring Boot Admin

Depending on how many Actuator endpoints are enabled in that application, you will see either more or fewer options in the left pane. As you activate the `health`, `env`, `metrics`, `beans`, and `loggers` endpoints, you will see **Details**, **Metrics**, **Environment**, **Beans**, and **Loggers**. If you open **Loggers**, you will see all loggers defined by the application. As you did in the *Changing settings in a running application* recipe in *Chapter 3*, you can change the log level, but this time from a nice UI:

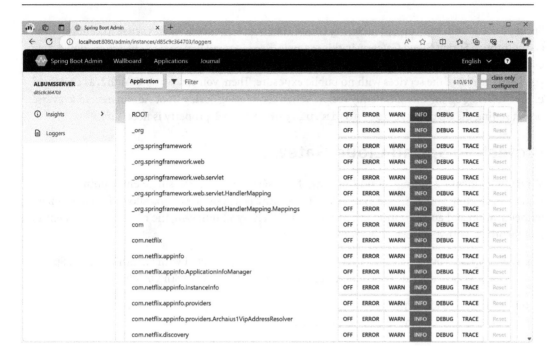

Figure 4.11: Loggers in Spring Boot Admin

There are two more views on the top bar:

- **Wallboard**: This shows the applications running in the wallboard view
- **Journal**: This shows the events that are happening in the Discovery service

How it works...

Spring Boot Admin may work without Eureka Server, but you would need to configure each application as a client of Spring Boot Admin. Instead, we configured Spring Boot Admin to discover the applications using Eureka Server. Connecting to Eureka Server requires Eureka Client. The Config service centralizes the configuration, which is why we used the Config Client.

Spring Boot Admin gets the list of applications and their instances from Eureka Server. Then, using the Actuator endpoint of each instance, it can get all the details of the application. The more Actuator endpoints are enabled, the more details can be shown. We used the central configuration to allow the desired endpoints in one single place.

Spring Boot Admin can run out of Spring Cloud Gateway; however, it makes sense to centralize the access through Spring Cloud Gateway in this example. Keep in mind that some Actuator endpoints may expose sensitive information. With this design, you only need to expose Spring Cloud Gateway while you keep the rest of the services with no public exposure. Then, you can set up OAuth2, as explained in the *Protecting Spring Cloud Gateway* recipe. When configuring Spring Boot Admin behind a reverse proxy, setting the `spring.boot.admin.ui.public-url` property is necessary.

Protecting Spring Cloud Gateway

When implementing Spring Cloud Gateway, it can serve as a system's single entry point. For this reason, protecting Spring Cloud Gateway with OAuth2 is a good idea. This allows for centralizing authentication and authorization in Spring Cloud Gateway, eliminating the need for your client to reauthenticate with each service behind it.

You want to place your `football` RESTful API, which is protected with OAuth2, behind Spring Cloud Gateway. So, you'll also need to protect Spring Cloud Gateway with OAuth2.

In this recipe, you'll learn how to configure Spring Cloud Gateway as a resource server and pass the token that you receive to the downstream service.

Getting ready

In this exercise, you will need the following:

- An authorization server. You can reuse Spring Authorization Server, which you created in the *Setting up Spring Authorization Server* recipe in *Chapter 2*, for this purpose.

- A resource server. The RESTful API you created in the *Protecting a RESTful API using OAuth2* recipe in *Chapter 2*, can be reused here.

- A Spring Cloud Gateway server. You can reuse the Spring Cloud Gateway server you created in the *Setting up Spring Cloud Gateway* recipe. You can always reuse the latest version of the Spring Cloud Gateway server in later recipes. I'm using the initial setup for simplicity.

- Eureka Server. You can reuse the Eureka Server application you created in the *Setting up Eureka Server* recipe.

If you haven't completed the previous recipes yet, I've prepared a working version for all of them in this book's GitHub repository at `https://github.com/PacktPublishing/Spring-Boot-3.0-Cookbook`, in the `chapter4/recipe4-9/start` folder.

How to do it...

In this recipe, we'll set our RESTful API behind Spring Cloud Gateway and then protect Spring Cloud Gateway with OAuth2. Let's begin:

1. First, configure the RESTful API so that it's registered in Eureka Server. For that, add the Eureka Client dependency to the RESTful API's pom.xml file:

```
<dependency>
    <groupId>org.springframework.cloud</groupId>
    <artifactId>spring-cloud-starter-netflix-eureka-client</
artifactId>
</dependency>
```

As it is part of Spring Cloud, you should also include the corresponding dependency management in the pom.xml file, as follows:

```
<dependencyManagement>
    <dependencies>
        <dependency>
            <groupId>org.springframework.cloud</groupId>
            <artifactId>spring-cloud-dependencies</artifactId>
            <version>${spring-cloud.version}</version>
            <type>pom</type>
            <scope>import</scope>
        </dependency>
    </dependencies>
</dependencyManagement>
```

Add a project-level property to configure the Spring Cloud version:

```
<properties>
    <spring-cloud.version>2022.0.4</spring-cloud.version>
</properties>
```

Now, you can add the Eureka Server configuration to the application.yml file:

```
eureka:
  client:
    serviceUrl:
      defaultZone: http://localhost:8761/eureka/
```

Though not required, I recommend configuring the application port randomly and assigning a name to the Spring Boot application. With this configuration, you won't need to care about port conflicts, and you'll make the application discoverable by name. For that, in the application.yml file, add the following lines:

```
spring:
  application:
```

```
        name: football-api
     server:
       port: 0
```

I've added the `spring` label for clarity, but the `application.yml` file should have it defined.

2. Next, configure Spring Cloud Gateway as a resource server. For that, you will need to add the Spring OAuth2 Resource Server dependency to your `pom.xml` file:

```
<dependency>
    <groupId>org.springframework.boot</groupId>
    <artifactId>spring-boot-starter-oauth2-resource-server</artifactId>
</dependency>
```

Then, configure `application.yml` with the application registration settings. We'll use the same configuration that we did for the RESTful API:

```
spring
  security:
    oauth2:
      resourceserver:
        jwt:
          audiences:
          - football
          - football-ui
          issuer-uri: http://localhost:9000
```

3. Now, configure Spring Cloud Gateway with the route to the RESTful API:

```
spring:
  cloud:
    gateway:
      routes:
        - id: teams
          uri: lb://football-api
          predicates:
            - Path=/football/**
```

I've included the `spring` field in this code snippet for clarity, but it's already defined for the security configuration, so you don't need to include it again; only copy the configuration that belongs to the `cloud` label.

4. Now that the application is behind Spring Cloud Gateway, which is protected using OAuth2, you can test the application. Remember to run the Eureka and Authorization projects before running the Spring Cloud Gateway and RESTful API projects.

First, you'll need to obtain an access token from the authorization server to test the application. For that, execute the following command in your Terminal:

```
curl --location 'http://localhost:9000/oauth2/token' \
--header 'Content-Type: application/x-www-form-urlencoded' \
--data-urlencode 'grant_type=client_credentials' --data-
urlencode 'client_id=football' \
--data-urlencode 'client_secret=SuperSecret' --data-urlencode
'scope=football:read'
```

Get the access token value and include and pass it in the authorization header to get the results through the application gateway. Spring Cloud Gateway listens on port 8080, so you need to execute the following request:

```
curl --location http://localhost:8080/football/teams -H
"Authorization: Bearer <access token>"
```

Replace <access token> with the access token you obtained from the authorization server.

You will see the result that's returned by the RESTful API:

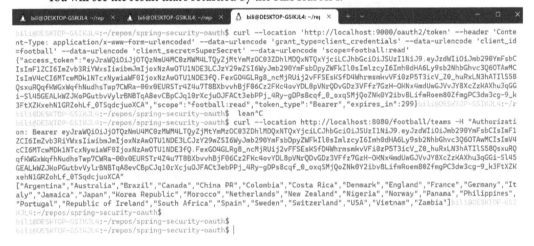

Figure 4.12: Using the RESTful API through Spring Cloud Gateway, which is protected with OAuth2

The result contains a list of teams.

How it works...

Spring Cloud Gateway acts as a resource server. This means it will require a valid access token to be issued by our authorization server.

Spring Cloud Gateway will relay the access token to the downstream RESTful API. Both will validate the access token. You can configure Spring Cloud Gateway with the first level of OAuth2 validation. For example, you can validate the token issuer and the token scopes. Then, if you need more fine-grained validation, you can do so on the RESTful API.

Part 2:
Database Technologies

Almost all applications need to persist and access data efficiently, and for that, Spring Boot offers many choices, from relational and NoSQL databases to repositories, templates, **Java Persistence Query Language (JPQL)**, and native SQL.

This part has the following chapters:

- *Chapter 5, Data Persistence and Relational Database Integration with Spring Data*
- *Chapter 6, Data Persistence and NoSQL Database Integration with Spring Data*

Data Persistence and Relational Database Integration with Spring Data

Most applications handle their data in some way, necessitating the use of a database engine. This chapter discusses Relational Databases, the most widely used database technology. Relational databases remain a flexible and dependable option for a variety of application scenarios. Their organized, tabular data storage format with a defined schema suits many purposes. Additionally, relational databases offer essential benefits like enforcing data integrity, supporting complex queries, and adhering to **ACID** principles (**Atomicity, Consistency, Isolation, Durability**). They prove to be a suitable choice for applications ranging from simple to mission-critical ones.

Spring Data is a component of the Spring Framework, designed to streamline data access in Java applications. It offers a consistent programming model and an abstraction layer for interacting with various data stores, including relational databases and other types of databases.

There are two modules in Spring Data for relational databases: Spring Data JPA and Spring Data JDBC.

- Spring Data JPA. This module provides integration with **Java Persistence API (JPA)**, allowing developers to work with relational databases using **Object-Relational Mapping** principles (**ORM**). One of the benefits is that most of the code is database independent, not for the purpose of creating an application totally independent of the database, but to reuse the learnings regardless the underlaying database. In complex applications, taking advantage of vendor-specific features can be determinant for a successful project, so I recommend using all features available from a database engine. Trying to make an application that can be deployed in any database causes that your application will use only the minimum common set available in all databases.

- Spring Data JDBC. This module offers more direct approach to database access, focusing on the use of plain SQL queries and direct mapping of data between Java objects and database tables.

We will use Spring Data JPA for the most common data access scenarios. From basic data operations such as Create, Read, Update, Delete (CRUD) to more advanced tasks such as complex queries, transactions, and database schema initialization and schema upgrade.

We will use PostgreSQL as a database engine, as it is Open Source, widely adopted, multiplatform and has a vibrant community around it. But as mentioned above, we could use the same principles to create an application using another relational database engine, such as MySQL, SQL Server or Oracle.

In this chapter we are going to cover the following main topics:

- Connecting your application to Postgresql
- Creating and updating the database schema
- Creating a CRUD repository.
- Using JPQL.
- Using Native queries.
- Updating operations
- Dynamic Queries
- Using Transactions
- Using Spring Data JDBC.

Technical requirements

For this chapter, you will need a PostgreSQL server. The easiest way to deploy it in your local environment is by using Docker. You can get Docker from the product page: `https://www.docker.com/products/docker-desktop/`

If you prefer to install PostgreSQL in your computer, you can download it from the project page: `https://www.postgresql.org/download/`

I also recommend installing PgAdmin to access the database. You can use it to observe the changes performed by your application in the database. You can download it from the project page: `https://www.pgadmin.org/download/`

You can use other tools, such as plugins for Visual Studio Code or IntelliJ.

You will need a code editor and OpenJDK as explained in the previous chapter.

All the recipes that will be demonstrated in this chapter can be found at: `https://github.com/PacktPublishing/Spring-Boot-3.0-Cookbook/tree/main/chapter5`.

Connecting your application to PostgreSQL

You want to create a RESTful API to server Football data to your end users. To manage this data, we decided to use a relational database, as we are interested in prodiving data consistency and advanced query capabilities.

In this recipe, we will connect an application, a RESTful API, to a PostgreSQL database. To do that, the first thing we'll do is deploy a PostgreSQL database in Docker.

In this recipe you will learn how to create a basic application that connects to a PostgreSQL database and perform basic SQL queries with JdbcTemplate.

Getting ready

For this recipe, you will need a PostgreSQL database. If you have already one server available, you can use it. Otherwise, you can use Docker to deploy a PostgreSQL in your computer. For that, you can execute the following command in your terminal to download and execute a PostgreSQL instance:

```
docker run -itd -e POSTGRES_USER=packt -e POSTGRES_PASSWORD=packt -p
5432:5432 --name postgresql postgres
```

You will have a PostgreSQL server available listening on port 5432, with username and password *packt*. If you want to change these parameters, you can modify the command above.

You will need a tool to perform some actions on PostgreSQL. I will use the command line tool psql. In Ubuntu you can install it using *apt,* the default package manager:

```
sudo apt install postgresql-client
```

Alternativelly to psql, you can use PgAdmin to connect to the database with a nice UI. I'll explain the samples just with the command line using psql, but you can use PgAdmin to execute the database scripts if you want. Follow the instruction from the official page at `https://www.pgadmin.org/download/` to install it in your computer.

You will find the sql scripts in the book's GitHub repository at `https://github.com/PacktPublishing/Spring-Boot-3.0-Cookbook/`.

As usual, we will use Spring Initializr tool to create our projects, or the integrated tool in your favorite IDE or editor if you prefer.

How to do it...

Once we have our PostgreSQL server ready as explained in *Getting Ready*, we will create a database. After this, we will create a Spring Boot application that will connect to the database to perform a simple query.

1. First, download the postgresql scripts available in the GitHub repository. They are located in `chapter5/recipe5-1/start/sql`. There are two script files:

   - `db-creation.sql`. This script creates a database named `football`, with two tables: `teams` and `players`.

   - `insert-data.sql`. This script inserts sample data in the `teams` and `players` tables.

2. Next, we'll execute the scripts in the database to execute the scripts. For that, open a terminal and execute the following commands to execute the scripts in PostgreSQL using the *psql* tool.

   ```
   psql -h localhost -U packt -f db-creation.sql
   psql -h localhost -U packt -f insert-data.sql
   ```

   It will request the password. Introduce *packt* as configured in the *Getting Ready* section. If you used different parameters in the *Getting Ready* section, use it accordingly.

   Alternatively, you can use *PgAdmin* tool instead of *psql* tool.

3. The database schema we just created looks like this:

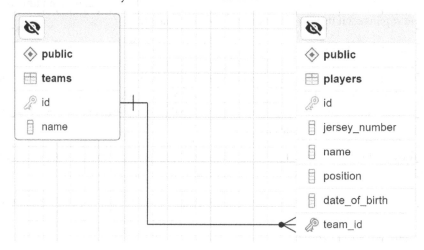

Figure 5.1: Database schema. Exported with PgAdmin tool.

4. Let's create a new Spring Boot application connecting to the database using *Spring Initializr* tool. We'll use the same parameters as in *Creating a RESTful API* recipe in *Chapter 1*, except changing the following options:

    - **Artifact**: `footballpg`

    - **Dependencies**: **Spring Web**, **Spring Data JPA**, **PostgreSQL Driver**

5. Next, we'll configure the application to connect to the PostgreSQL database. For that, create an `application.yml` file in the `resources` folder and set the following content:

    ```yaml
    spring:
        datasource:
            url: jdbc:postgresql://localhost:5432/football
            username: packt
            password: packt
    ```

6. Now, create a new service class named `TeamsService`. This class will use JdbcTemplate to perform the queries against the database. For that, it's necessary to inject a JdbcTemplate.

    ```java
    @Service
    public class TeamsService {
        private JdbcTemplate jdbcTemplate;
        public TeamsService(JdbcTemplate jdbcTemplate) {
            this.jdbcTemplate = jdbcTemplate;
        }
    }
    ```

7. We'll create a method in the *TeamsService* that gets the total number of teams. You can name the method `getTeamCount`:

    ```java
    public int getTeamCount() {
        return jdbcTemplate.queryForObject("SELECT COUNT(*) FROM
    teams", Integer.class);
    }
    ```

    We used the jdbcTemplate method queryForObject to perform an SQL query to get a single value.

8. You can now create a RestController using this service. I made a sample controller that uses the `TeamsService`. You can find it in the book's repository at `https://github.com/PacktPublishing/Spring-Boot-3.0-Cookbook/`.

How it works...

In our configuration file `application.yml`, we have defined a data source. Within the data source, we have used a URL, a username, and a password. However, it is also possible to define additional properties. Alternatively, we can use only the URL. The URL contains important information such as the type of the database (in our case, PostgreSQL), the host, the port, and the database name. Although it's possible to pass the username and password in the URL, we used specific fields to enhance clarity.

Since we specified PostgreSQL as the database, it is essential to have a driver registered in the class path. We achieved that by adding the dependency to the PostgreSQL driver.

A data source object is registered in the dependency container by defining a data source in the configuration file. Spring Data JPA uses that data source to create JdbcTemplates when needed, for instance, it creates a JdbcTemplate instance and injects it into the `TeamsService` class.

JdbcTemplate handles the creation and release of resources and converts `SQLExceptions` into Spring's `DataAccessExceptions`. In this example, we used a very simple query that does not require any parameter and returns an Integer. Still, JdbcTemplate allows passing parameters to your query and mapping the results to complex classes. We won't extend much on these capabilities in this book; rather, we'll go deeper into JPA and Hibernate capabilities to map complex entities and relations to classes. We'll see that in from *Using Hibernate* recipe onwards.

There's more...

JdbcTemplate can be used to retrieve results that are not limited to scalar values. For instance, assuming that we have a `Team` class, we can define the following method that uses the `query` method to retrieve all teams:

```
public List<Team> getTeams() {
    return jdbcTemplate.query("SELECT * FROM teams", (rs, rowNum) -> {
        Team team = new Team();
        team.setId(rs.getInt("id"));
        team.setName(rs.getString("name"));
        return team;
    });
}
```

In this example, we use an anonymous `RowMapper` that transforms each row into a `Team`.

You can also pass arguments to your query. For instance, let's retrieve a specific team:

```
public Team getTeam(int id) {
    return jdbcTemplate.queryForObject(
        "SELECT * FROM teams WHERE id = ?",
        new BeanPropertyRowMapper<>(Team.class),
```

```
        id);
}
```

This time, we used a `BeanPropertyRowMapper` to map the resulting row to Team. This class infers the target properties to map the columns of the resulting rows.

Using JdbcClient to access the database

In the previous recipe, we used JdbcTemplate to access the database. The JdbcClient is an enhanced JDBC client that provides a Fluent interaction mode. JdbcClient has been introduced in Spring Framework 6.1, and it's available since Spring Boot 3.2.

In this recipe, we'll learn how to use JdbcClient by performing some simple queries to the database.

Getting ready

In this recipe, we'll need a PostgreSQL database. You can reuse the same database created in the previous recipe, *Connecting your application to PostgreSQL*. You can reuse the project from the same recipe as well, as the dependencies are the same. I prepared a working version that you can use as a starting point for this recipe. You can find it in the book's GitHub repository at `https://github.com/PacktPublishing/Spring-Boot-3.0-Cookbook`, in `chapter5/recipe5-2/start` folder.

How to do it...

Let's prepare some queries using JdbcClient instead of JdbcTemplate.

1. Let's start by creating a new service class named `PlayersService` and inject a JdbcClient in the constructor:

    ```
    @Service
    public class PlayersService {
        private JdbcClient jdbcClient;

        public PlayersService(JdbcClient jdbcClient) {
            this.jdbcClient = jdbcClient;
        }
    }
    ```

2. Create a class named `Player`. This class should have the same fields as the table `players` created in the *Connecting your application to PostgreSQL* recipe. You can find an implementation of this class in the book's repository at `https://github.com/PacktPublishing/Spring-Boot-3.0-Cookbook`.

3. Now, we can create methods in the `PlayersService` to interact with the database:

    - Let's create a method named `getPlayers` to retrieve all players:

    ```
    public List<Player> getPlayers() {
        return jdbcClient.sql("SELECT * FROM players")
                .query(Player.class)
                .list();
    }
    ```

    - We can create a method named `getPlayer` to retrieve a single `Player`. We can use a parameter in the SQL query.

    ```
    Public Player getPlayer(int id) {
        return jdbcClient.sql("SELECT * FROM players WHERE id =
    :id")
                .param("id", id)
                .query(Player.class)
                .single();
    }
    ```

    - Let's do a method to create a new `Player`. Name it `createPlayer`:

    ```
    public Player createPlayer(Player player) {
        GeneratedKeyHolder keyHolder = new GeneratedKeyHolder();
        jdbcClient.sql("""
          INSERT INTO players (jersey_number, name, position, date_
    of_birth, team_id)
          VALUES (:jersey_number, :name, :position, :date_of_birth,
    :team_id)
                        """)
                .param("name", player.getName())
                .param("jersey_number", player.getJerseyNumber())
                .param("position", player.getPosition())
                .param("date_of_birth", player.getDateOfBirth())
                .param("team_id", player.getTeamId())
                .update(keyHolder, "id");
        player.setId(keyHolder.getKey().intValue());
        return player;
    }
    ```

4. You can create a controller that uses the `PlayerService`. I prepared a working version that you can find in the book's GitHub repository at: `https://github.com/PacktPublishing/Spring-Boot-3.0-Cookbook`.

How it works...

The mechanism used by the JdbcClient to create a connection to the database is like the JdbcTemplate. Spring Data JPA uses the data source configured in the application and injects it in the JdbcClient.

The JdbcClient provides a Fluent way to interact with the database, making the development more intuitive and reducing the boilerplate code. It allows using named parameters very easily, as we saw in the methods `getPlayer` and `createPlayer`. It also provides automatic mapping, with no need to define a `RowMapper` to process each row.

Using an ORM to access the database

Accessing a database performing SQL requests can be performant and can work for simple applications. However, when the application becomes more complex and the database schema grows, it can be interesting using an **Object-Relational Mapping (ORM)** framework to access the database using an **Object-Oriented Programming (OOP)** layer. Probably, the most popular ORM framework in Java is **Hibernate**, and Spring Data JPA uses Hibernate as its default **Java Persistence API (JPA)** implementation.

In this recipe, we will create entity classes that can be mapped to the database schema, and we'll interact with the database without writing a single line of SQL.

Getting ready

You will need a PostgreSQL database for this recipe. You can reuse the database created in the *Connecting your application to PostgreSQL* recipe. If you haven't completed that recipe yet, you can complete the first two steps of that recipe to create the database.

How to do it...

Once we have our PostgreSQL server ready, as explained in *Getting Ready*, we will create a database, and we will fill it with some data. After this, we will create a RESTfull project to connect to the database to retrieve the data.

1. Let's create a project connected to this database. We will use Spring Initialzr tool by opening `https://start.spring.io` in your browser.

    We will use the usual parameters as in *Chapter 1*, with the following specific parameters:

    • **Artifact**: `footballpg`

    • **Dependencies**: **Spring Web, Spring Data JPA, PostgreSQL Driver**

2. In this project, create the entity classes to map with the database tables.

- Create a file named TeamEntity.java containing a class to map with table teams:

```java
@Table(name = "teams")
@Entity
public class TeamEntity {
    @Id
    private Integer id;
    private String name;
    @OneToMany(cascade = CascadeType.ALL, mappedBy = "team")
    private List<PlayerEntity> players;
}
```

- Create a file named PlayerEntity.java containing the class to map with table players.

```java
@Table(name = "players")
@Entity
public class PlayerEntity {
    @Id
    private Integer id;
    private Integer jerseyNumber;
    private String name;
    private String position;
    private LocalDate dateOfBirth;

    @ManyToOne(fetch = FetchType.LAZY)
    @JoinColumn(name = "team_id")
    private TeamEntity team;
}
```

3. Create two repository interfaces to access the database using the entities we created in the previous step.

- Create a file named TeamRepository.java with the following interface:

```java
public interface TeamRepository extends
CrudRepository<TeamEntity, Integer>{
}
```

- Create a file named PlayerRepository.java containing the following interface:

```java
public interface PlayerRepository extends
JpaRepository<PlayerEntity, Integer>{
    List<PlayerEntity> findByDateOfBirth(LocalDate dateOfBirth);
    List<PlayerEntity> findByNameContaining(String name);
}
```

4. Create a service class named `FootballService` using both repositories:

```
@Service
public class FootballService {
    private PlayerRepository playerRepository;
    private TeamRepository teamRepository;

    public FootballService(PlayerRepository playerRepository,
TeamRepository teamRepository) {
        this.playerRepository = playerRepository;
        this.teamRepository = teamRepository;
    }

}
```

5. Create two classes representing the data exposed:

- Team:

```
public record Team(Integer id, String name, List<Player>
players) {
}
```

- Player:

```
public record Player(String name, Integer jerseyNumber, String
position, LocalDate dateOfBirth) {
}
```

6. Add a couple of methods to find players using different criteria:

- Search players that contain a given string in the name:

```
public List<Player> searchPlayers(String name) {
        return playerRepository.findByNameContaining(name)
                .stream()
                .map(player -> new Player(player.getName(),
player.getJerseyNumber(), player.getPosition(), player.
getDateOfBirth()))
                .toList();
    }
```

- Search players by birth date:

```
public List<Player> searchPlayersByBirthDate(LocalDate date) {
        return playerRepository.findByDateOfBirth(date)
                .stream()
                .map(player -> new Player(player.getName(),
player.getJerseyNumber(), player.getPosition(), player.
```

```
            getDateOfBirth()))
                          .toList();
        }
```

7. Add a method to return a Team, including its players:

```
@Transactional(readOnly=true)
public Team getTeam(Integer id) {
    TeamEntity team = teamRepository.findById(id).orElse(null);
    if (team == null) {
        return null;
    } else {
        return new Team(team.getId(),
                team.getName(),
                team.getPlayers()
                .stream()
                .map(player -> new Player(player.getName(),
player.getJerseyNumber(), player.getPosition(),
                                    player.getDateOfBirth())))
                .toList());
    }
}
```

8. Create a new team:

```
public Team createTeam(String name) {
    Random random = new Random();
    TeamEntity team = new TeamEntity();
    Integer randomId = random.nextInt();
    if (randomId < 0) {
            randomId = random.nextInt();
    }
    team.setId(randomId);
    team.setName(name);
    team = teamRepository.save(team);
    return new Team(team.getId(), team.getName(), List.of());
}
```

9. Update the position of a player:

```
public Player updatePlayerPosition(Integer id, String position)
{
    PlayerEntity player = playerRepository.findById(id).
orElse(null);
    if (player == null) {
        return null;
```

```
    } else {
        player.setPosition(position);
        player = playerRepository.save(player);
        return new Player(player.getName(), player.
getJerseyNumber(), player.getPosition(),
                    player.getDateOfBirth());
    }
}
```

10. Now, you can create a controller using the service to expose the application's logic. You can find a complete example in the book's GitHub repository at `https://github.com/PacktPublishing/Spring-Boot-3.0-Cookbook`.

11. Now configure the application to connect to PostgreSQL database. Under `resources` folder, create a file named `application.yml`. Set the following configuration:

```
spring:
    jpa:
        database-platform: org.hibernate.dialect.
PostgreSQLDialect
        open-in-view: false
    datasource:
        url: jdbc:postgresql://localhost:5432/football
        username: packtd
        password: packtd
```

12. Now you can execute and test the application. You can use the instructions from *Chapter 1* to test the application using curl. I also provided a script that you can find in the repository with curl requests for this application. It is located at `chapter2/recipe2-1/end/scripts/requests.sh`

How it works...

Hibernate is an **Object-Relational Mapping (ORM)** framework. Its primary goal is to bridge the gap between the Java programming language and relational databases. Using annotations such as @Entity, @Table, @Id, @OneToMany, @ManyToOne, and others not used in this recipe, Hibernate maps the classes to database tables. These mapped classes are known as *Entities*. Hibernate also provides other features, such as transaction management, query capabilities, caching, and lazy loading.

Hibernate is the default JPA provider for Spring Data JPA. Spring Data JPA allows you to define Repository interfaces to interact with your data model. Just by extending the `CrudRepository` interface, it automatically generates the necessary JPA operations to provide Create, Read, Update, and Delete operations for your Entities. When using `JpaRepository`, Spring Data JPA generates the necessary JPA queries based on method names.

For instance, we used `findByDateOfBirth` to create a method that returns all players by their date of birth and `findByNameContaining` to return all players whose names contain a given string. All this without writing a single line of SQL!

If you are unfamiliar with the naming convention, I strongly recommend checking the project documentation. See `https://docs.spring.io/spring-data/jpa/ reference/#repository-query-keywords`.

In addition to the operations to read data, `CrudRepository` and `JpaRepository` have a method named `save`. This method allows you to update existing entities or create new ones if they don't already exist. They also have some methods to delete entities, such as `delete`, `deleteById` and other methods.

Even with the abstraction offered by Spring Data JPA, it's crucial to understand some of the inner workings of Spring Data. In this recipe, I used a middle tier component named `FootballService` between the controller and the repositories. You could also call a repository directly from your controller, however there are some caveats with this approach. To better understand it, lets deep dive on the operation to return a team and its players.

Hibernate has two ways of loading the entities with related entities: **Lazy** and **Eager**. Lazy means that Hibernate will retrieve a related entity just when the member mapping the related entity is used. Eager, the other way, will retrieve the related data at the same time than the main entity is retrieved. Let's see this behavior with an example. The `TeamEntity` has a member annotated with @OneToMany to manage its Player Entities. When your application uses the method `getPlayers` of the class `TeamEntity`, Hibernate tries to load the players by performing a request to the database. By default, @OneToMany relations are loaded in Lazy mode, while @ManyToOne are loaded in Eager mode. In summary, Lazy means, that if you don't use the relation, the request to the database won't happen. If you use the repository to retrieve the team in the controller and you try to return the `TeamEntity`, it will serialize the entity into a Json object, traversing all properties, including the players. At that stage of the request execution, there is no session open to access the database and you will receive an exception. There are several ways to fix this issue:

- By retrieving the players at the same time as the team by using the Eager mode. It can be valid in certain scenarios, but it could cause unnecessary requests to the database.

- By allowing opening connections in view. This can be done using `spring.jpa.open-in-view=true`. This is an antipattern and I strongly recommend not using it.

  I had a bad experience related to this antipattern working on a project. I had an issue related to bad performance and availability of the application, but the system had resources and any component seemed under stress. Finally, I realized there were connection leaks in the application due to this `open-in-view` option. Finding the root cause and resolving this bug was one of the most challenging tasks I've encountered, as figuring out the root cause was not obvius.

  By the way, `spring.jpa.open-in-view=true` is the default value in Spring Boot, so keep this in mind and configure as false unless you have a good reason that I cannot imagine.

- By creating a session or a transaction while you are retrieving the data, including the lazy relations. This is the approach followed in this recipe. When mapping from `TeamEntity` to `Team`, we used the method `getPlayers`, and hence retrieved the data from the database. As the method `getTeam` in the `FootballService` is marked as `@Transactional`, all requests happen in the same transaction/session. As this is a read operation, you can set `@Transactional(readOnly = true)`, for having a less costly isolation mode for your transaction.

- By performing a `Join` query to retrieve both the team and its players in a single database request. This is the most efficient way to implement this scenario. We will see how to do it in another recipe of this chapter.

That is the reason to create a `Service` class and not return Entities directly in the RESTful API.

There's more...

In this exercise, we used an existing database, and we manually created the entities and repositories to interact with the database. There is another approach that we will tackle in further recipes of this chapter, that defines the entities first and then generates the database automatically. For both scenarios there are tools that assist you in this task that can be very mechanical. For instance, you can use JPA Buddy plugin for IntelliJ, `https://plugins.jetbrains.com/plugin/15075-jpa-buddy`. It has a basic free version that is enough for simple scenarios, and a paid one with advanced features.

In this recipe, we created some code to transform Entities into other objects, also known as **Data Transfer Objects** (**DTOs**). It could add a lot of boilerplate code in your project. There are libraries that automate mapping between Java Beans that can fit well in this scenario. For example Mapstruct (`https://mapstruct.org/`). Spring Data JPA supports using Mapstruct to convert Entities to DTO and viceversa. For learning purposes, I don't use it in the recipes.

See also

If you want to learn more about **Open Session In View** (**OSIV**) antipattern, I recommend you to read this article `https://vladmihalcea.com/the-open-session-in-view-anti-pattern/` to understand the details.

Creating the database schema from our code

Creating a database schema and its corresponding Entities in our application, as seen in the previous recipe, requires a lot of repetitive work. Instead, we can create our Entities and we can generate the database schema automatically. In this recipe, we will use Spring Data JPA to generate the database schema based on the entity model of the application.

Getting ready

For this recipe, you will need the same tools as in the previous recipe, namely a PostgreSQL server that you can run in a Docker container or on your computer.

We will use the same code generated in the previous recipe. If you didn't complete it, you can find a completed recipe in the book's GitHub repository https://github.com/PacktPublishing/Spring-Boot-3.0-Cookbook/.

How to do it...

We will use the previous example about football teams and players. Still, in this recipe, instead of using an existing database and creating the mapping entities, we will go in the opposite direction. We will use and tune the Entities already created to generate the Database schema. Let's begin:

1. Create a new database named football2.

    *   Open psql in your terminal.

        ```
        psql -h localhost -U packtd
        ```

    *   Execute the following SQL command to create the database:

        ```
        CREATE DATABASE football2;
        ```

2. Instead of manually generating the identifiers of our entities we will rely on the automatic identifier generators of the database. For that purpose, we will modify the annotation @Id of our entities.

    *   Open TeamEntity and modify the member id as follows:

        ```
        @Id
        @GeneratedValue(strategy = GenerationType.IDENTITY)
        private Integer id;
        ```

    *   Do the same with PlayerEntity.

3. Open application.yml file and add the spring.jpa.generate-ddl=true and spring.sql.init.mode=always properties. The file should look like this:

        ```
        spring:
          jpa:
              database-platform: org.hibernate.dialect.
        PostgreSQLDialect
              open-in-view: false
              generate-ddl: true
          sql:
        ```

```
        init:
            mode: always
    datasource:
        url: jdbc:postgresql://localhost:5432/football2
        username: packtd
        password: packtd
```

If you run the application, the database schema will be created automatically.

4. Modify the `createTeam` method in class `FootballService`:

```
public Team createTeam(String name) {
    TeamEntity team = new TeamEntity();
    team.setName(name);
    team = teamRepository.save(team);
    return new Team(team.getId(), team.getName(), List.of());
}
```

Here, we removed the generation of the team identifier, instead it will be automatically generated.

5. Copy the file located in the GitHub repository, in `chapter5/recipe5-4/start/data.sql`, to the resources folder. This folder is located in `src/main/resources`.

6. Execute the application.

7. Test the application by performing requests to the application as explained in *Chapter 1*. I also provided a script that you can find in the repository with curl requests for this application. It is located at `chapter5/recipe5-4/end/scripts/requests.sh`.

    You will see that the database schema is initialized, and it already has data.

How it works...

By configuring the application with *spring.jpa.generate-ddl=true*, Spring Data will automatically generate the data schema from the entities defined in the project. It will use the annotations to generate the schema according to the target database. For instance, we used `@GeneratedValue` for id field in both `PlayerEntity` and `TableEntity`. It is translated into a PostgreSQL sequences. Taking the `TeamEntity` as an example, this is the result in PostgreSQL:

```
CREATE TABLE IF NOT EXISTS public.teams
(
    id integer NOT NULL DEFAULT nextval('teams_id_seq'::regclass),
    name character varying(255) COLLATE pg_catalog."default",
    CONSTRAINT teams_pkey PRIMARY KEY (id)
)
CREATE SEQUENCE IF NOT EXISTS public.teams_id_seq
    INCREMENT 1
    START 1
```

```
MINVALUE 1
MAXVALUE 2147483647
CACHE 1
OWNED BY teams.id;
```

Spring Boot is able to create the schema and initialize data. It loads schema scripts from `optional:classpath*:schema.sql` and data scripts from `optional:classpath*:data.sql`. We only provided data scripts explicitly, and we did let Spring Boot generate the schema with `generate-ddl` setting. In addition to the data scripts, you can also provide the schema scripts instead of letting Spring Boot generate them for you. For complex applications, probably you will require specific database settings..

As mentioned, in this recipe we let Spring Boot performing the database initialization. By default, Spring Boot only performs the initialization if it considers that the database is an In-Memory embedded database, such as H2. To force the initialization for PostgreSQL, we used the parameter `spring.sql.init.mode=always`.

The approach followed in this recipe is intended for development environments. In a production environment, we could have multiple instances of the same application and it can cause issues having more than one instance trying to initialize the database. Even with mechanisms in place to ensure that only one instance of the application is updating the database, this process can take time and will slow down the application initialization. It is important to note, that some of those scripts should be executed only once. For instance, in this recipe we used a `data.sql` that insert records in both tables using explicit id values. If you try to execute it twice will produce a unique constraint validation error. For initialization you most probably want to execute the process before all the application instances start. For instance, in Kubernetes you can achieve this by using Init Containers, see `https://kubernetes.io/docs/concepts/workloads/pods/init-containers/`.

For production environments, other tools such as Flyway and Liquibase exist, and they are supported by Spring Boot. Those tools provide more control on the database creation, providing versioning and migrations. In the next recipe, we'll use Flyway to create and migrate the schema of the database.

There's more...

In this recipe, we used few options among all possibilities to customize our entities, but almost any aspect of database schema definition can be controlled. Just to give some examples:

- `@Entity`: Annotating a class with `@Entity` indicates that it is a JPA entity and should be mapped to a database table. Each entity class corresponds to a table in the database, and each field in the class corresponds to a column in the table.

- `@Table`: It is used to specify the details of the database table to which an entity should be mapped. You can use it to set the table name, schema, and other attributes.

- `@Column`: It allows you to configure the mapping of an entity field to a database column. You can specify attributes like column name, length, nullable, and unique constraints.

- `@JoinColumn`: It is used to specify the column that represents a foreign key in a relationship. It is often used in conjunction with `@ManyToOne` or `@OneToOne` to specify the join column's name and other attributes.

- `@Transient`: Fields marked with `@Transient` are not mapped to database columns. This annotation is used for fields that should be excluded from database persistence.

- `@Embedded` and `@Embeddable`: These annotations are used for creating embedded objects within entities. `@Embeddable` is applied to a class, and `@Embedded` is used in an entity to indicate that an instance of the embedded class should be persisted as part of the entity.

- `@Version`: It is used to specify a version property for optimistic locking. It is typically applied to a numeric or timestamp field and is used to prevent concurrent updates to the same record.

See also

`https://docs.spring.io/spring-boot/docs/current/reference/html/howto.html#howto.data-initialization`

PostgreSQL integration tests with Testcontainers

When developing tests for a component, one of the biggest challenges is managing dependent services like databases. While creating mocks or using an in-memory database like H2 can be a solution, these approaches may hide potential issues in our application. **Testcontainers** is an open-source framework that offers temporary instances of popular databases and other services that can be run on Docker containers. This provides a more reliable way to test applications.

In this recipe, you will learn how to create an integration test that depends on PostgreSQL using Testcontainers.

Getting ready

In this recipe, we'll create some tests for the application created in the previous recipe, *Creating the database schema from our code*. I prepared a working version as a starting point for this recipe in case you haven't completed the previous recipe yet. You can find it in the book's GitHub repository at `https://github.com/PacktPublishing/Spring-Boot-3.0-Cookbook`.

Testcontainers requires Docker installed in your computer.

How to do it...

Let's enhance the reliability of our application by creating tests that utilize a real PostgreSQL database.

1. First, we'll need to include the Testcontainers starter and the PostgreSQL Testcontainer dependency. You can do it by adding the following dependencies in the project pom.xml file:

```
<dependency>
    <groupId>org.testcontainers</groupId>
    <artifactId>junit-jupiter</artifactId>
    <scope>test</scope>
</dependency>
<dependency>
    <groupId>org.testcontainers</groupId>
    <artifactId>postgresql</artifactId>
    <scope>test</scope>
</dependency>
```

2. Next, create a test class, you can name it FootballServiceTest. Let's setup the class for TestContainers. For that we'll need:

    - Annotate the class with @SpringBootTest.

    - Annotate the class with @TestContainers.

    - Configure a context initializer that configures the application context using the PostgreSQL container that we'll create during the test. To set the initializer we can annotate the class with @ContextConfiguration.

    The class definition will look like this:

```
@SpringBootTest
@Testcontainers
@ContextConfiguration(initializers = FootballServiceTest.
Initializer.class)
public class FootballServiceTest
```

As you can see, there is a reference to FootballServiceTest.Initializer class that we haven't described yet. It's explained in the following steps.

3. Now, we'll define a static field with the PostgreSQL container:

```
static PostgreSQLContainer<?> postgreSQLContainer = new
PostgreSQLContainer<>("postgres:latest")
        .withDatabaseName("football")
        .withUsername("football")
        .withPassword("football");
```

4. Let's use the container we just created to configure the application. It's now when we'll create the `FootballServiceTest.Initializer` class. Inside the FootballServiceTest create a class named Initializer:

```
static class Initializer
            implements
ApplicationContextInitializer<ConfigurableApplicationContext> {
    public void initialize(ConfigurableApplicationContext
configurableApplicationContext) {
        TestPropertyValues.of(
                "spring.datasource.url=" +
postgreSQLContainer.getJdbcUrl(),
                "spring.datasource.username=" +
postgreSQLContainer.getUsername(),
                "spring.datasource.password=" +
postgreSQLContainer.getPassword())
        .applyTo(configurableApplicationContext.getEnvironment());
    }
}
```

The initializer overrides the data source configuration using the PostgreSQLContainer settings.

5. The final step in configuring Testcontainers is to start the container, which can be done before all tests begin using @BeforeAll annotation. Let's create a method that starts the container:

```
@BeforeAll
public static void startContainer() {
    postgreSQLContainer.start();
}
```

6. Now, we can create the tests normally. For example, let's make a test that creates a team:

```
@Autowired
FootballService footballService;
@Test
public void createTeamTest() {
    Team team = footballService.createTeam("Jamaica");
    assertThat(team, notNullValue());
    Team team2 = footballService.getTeam(team.id());
    assertThat(team2, notNullValue());
    assertThat(team2, is(team));
}
```

How it works...

The `@Testcontainers` annotation searches for all fields tagged with @Container and triggers their container lifecycle methods. Containers declared as static fields, like in this recipe, are shared between test methods. This means that the container is started only once, before any test method is executed, and stopped after the last test method has executed. If the container were declared as an instance field, it would be started and stopped for each test method.

The PostgreSQLContainer is a specialized Testcontainer module that exposes the attributes of the database to facilitate the connection in our tests. We used the JdbcUrl, the username and the password to override the configuration.

As you can see, we didn't require mocking any repository to create tests for the FootballService class. Another great advantage is that the database is recreated for every test execution cycle, so the tests are repeatable and predictable.

Versioning and upgrading database schema

As our applications evolve, we'll need to keep the database in sync with our Java entities. That could be a complex and error prone task. To address this scenario, there are tools to manage database schemas and database migrations. A couple of examples of such tools are Flyway and Liquibase, both supported by Spring Boot.

In addition to the database migration feature itself, Flyway provides the following features:

- Version control to keep track of the migrations applied to a database and the ones which are pending.

- It can be integrated into development environments and build automation tools, such as Maven or Gradle.

- Repeatable migrations. Every time Flyway runs, repeatable migrations are executed ensuring that the database remains in the desired state.

- Rollback and undo operations. Flyway can automatically generate SQL scripts to undo a specific migration, allowing rollbacks in case of issues.

- It can execute the migration during the initialization of your project.

- It provides standalone tools that can be used out of your Java project.

When used in your project, it requires different configurations to perform the migration, for instance registering specific beans. Spring Boot facilitates that integration, minimizing the necessary configuration to just some application settings unless you need more advanced actions.

Getting ready

You can use the exercise completed in the previous recipe as starting point for this recipe. If you haven't completed yet, you can find a complete version in the book's GitHub repository at `https://github.com/PacktPublishing/Spring-Boot-3.0-Cookbook/`. You will need the same tools, PostgreSQL and Docker.

Flyway is a solution maintained by Redgate, with free edition for individual and non-commercial projects and paid supported editions. For this recipe, we can use the libraries but keep in mind that using Flyway in a production environment may require a Redgate license. See `https://documentation.red-gate.com/fd/licensing-164167730.html` for details.

How to do it...

In this recipe, we'll use Flyway to create an initial version of the database, and then we'll apply changes. We'll learn how it can be easily used with Spring Boot.

1. Create a new database named `football3`.

   - Open psql in your terminal.

     ```
     psql -h localhost -U packtd
     ```

   - Execute the following SQL command to create the database:

     ```
     CREATE DATABASE football3;
     ```

2. Add Flyway dependency. In your `pom.xml` file add the dependency `org.flywaydb:flywaycore`.

   ```xml
   <dependency>
       <groupId>org.flywaydb</groupId>
       <artifactId>flyway-core</artifactId>
   </dependency>
   ```

3. Create the database creation script. The default location for flyway scripts is under `src/main/resources/db`. Name the file `V1_InitialDatabase.sql` and add the following content to create the `teams` and `players` tables:

   ```sql
   CREATE TABLE teams (
       id SERIAL PRIMARY KEY,
       name VARCHAR(255)
   );

   CREATE TABLE players (
       id SERIAL PRIMARY KEY,
   ```

```
        jersey_number INT,
        name VARCHAR(255),
        position VARCHAR(255),
        date_of_birth DATE,
        team_id INT REFERENCES teams(id)
);
```

You can also use this script to fill the database with data, for instance adding teams and players.

```
INSERT INTO teams(id, name) VALUES (1884881, 'Argentina');
INSERT INTO players(id, jersey_number, name, "position", date_
of_birth, team_id)
VALUES
(357669, 2, 'Adriana SACHS', 'Defender', '1993-12-25', 1884881)
```

In the book's GitHub repository, you can find more data for this script, you can copy and paste in your project if you wish.

Migration files naming convention

Migration files should follow the naming convention: V<version>__<name>.sql, version can be <major version>_<minor version>, but it is optional. Pay attention that between <version> and <name> there are two underscore signs.

---

4. Launch the application.

    If you look at the output logs, you will see a message similar to this:

```
main] o.f.c.i.s.JdbcTableSchemaHistory    : Schema history table "public"."flyway_schema_history" does not exist yet
main] o.f.core.internal.command.DbValidate : Successfully validated 1 migration (execution time 00:00.015s)
main] o.f.c.i.s.JdbcTableSchemaHistory    : Creating Schema History table "public"."flyway_schema_history" ...
main] o.f.core.internal.command.DbMigrate  : Current version of schema "public": << Empty Schema >>
main] o.f.core.internal.command.DbMigrate  : Migrating schema "public" to version "1 - InitialDatabase"
main] o.f.core.internal.command.DbMigrate  : Successfully applied 1 migration to schema "public", now at version v1 (execution time 00:00.128s)
```

Figure 5.2: Application logs showing migration execution.

Flyway created a new table to manage the schema history named flyway_schema_history and executed the script we create above. You can get the list of tables in PostgreSQL using command \dt.

```
packtd=# \c football3
psql (14.9 (Ubuntu 14.9-0ubuntu0.22.04.1), server 16.0 (Debian 16.0-1.pgdg120+1))
WARNING: psql major version 14, server major version 16.
        Some psql features might not work.
You are now connected to database "football3" as user "packtd".
football3=# \dt
               List of relations
  Schema |          Name           | Type  | Owner
 --------+-------------------------+-------+--------
  public | flyway_schema_history   | table | packtd
  public | players                 | table | packtd
  public | teams                   | table | packtd
 (3 rows)
```

Figure 5.3: List of tables in recently in the database created by Flyway.

The database now has the tables necessary to manage our application.

5. Let's create a migration for our application now. We need to manage football matches in our application, and we need to know the height and weight of the players.

  - The matches will be managed in a new Entity named `MatchEntity`. It will have two fields referencing the teams playing the match, the match date, and the goals scored by each team. It should look like this:

```
@Entity
@Table(name = "matches")
public class MatchEntity {
    @Id
    @GeneratedValue(strategy = GenerationType.IDENTITY)
    private Integer id;
    private LocalDate matchDate;
    @ManyToOne
    @JoinColumn(name = "team1_id", nullable = false)
    private TeamEntity team1;
    @ManyToOne
    @JoinColumn(name = "team2_id", nullable = false)
    private TeamEntity team2;
    @Column(name = "team1_goals", columnDefinition = "integer
default 0")
    private Integer team1Goals;
    @Column(name = "team2_goals", columnDefinition = "integer
default 0")
    private Integer team2Goals;
}
```

  - The existing entity `PlayerEntity` should have two new properties to manage the height and weight of the player.

```
private Integer height;
private Integer weight;
```

  - We need to create the SQL scripts for the database now. In `src/main/resources/db` create a new sql file named `V2__AddMatches.sql`. Add the required changes on the database to support the application.

  - Create the table `matches`

```
CREATE TABLE matches(
    id SERIAL PRIMARY KEY,
    match_date DATE,
    team1_id INT NOT NULL REFERENCES teams(id),
    team2_id INT NOT NULL REFERENCES teams(id),
```

```
        team1_goals INT default 0,
        team2_goals INT default 0
    );
```

- Modify the table players to add the two columns for the height and weight.

```
ALTER TABLE players ADD COLUMN height INT;
ALTER TABLE players ADD COLUMN weight INT;
```

- We can also set values for the existing players, for simplicity, we can set the same values for all players:

```
UPDATE players SET height = 175, weight = 70;
```

6. Execute the application.

   If you check the logs, you will see the schema migration applied by Flyway.

```
main] o.f.core.internal.command.DbValidate    : Successfully validated 2 migrations (execution time 00:00.020s)
main] o.f.core.internal.command.DbMigrate     : Current version of schema "public": 1
main] o.f.core.internal.command.DbMigrate     : Migrating schema "public" to version "2 - AddMatches"
main] o.f.core.internal.command.DbMigrate     : Successfully applied 1 migration to schema "public", now at version v2 (execution time 00:00.046s)
```

Figure 5.4: Application logs showing the new schema migration.

7. During the startup of the application, the database is initialized. In order to ensure that the migration is working as expected, you can use Testcontainers to validate it. You can check if certain data exists in the database using this approach. There are some tests available in the book's GitHub repository that assume the existence of certain values in the database.

How it works...

On adding the Flyway dependency to your project, it will check for migration scripts during the application startup. If there are migration scripts, it will connect to the database and check for migrations that have been already applied by looking at the table flyway_schema_history. If that table doesn't exist yet, it will create it. Then it will start executing all migrations that have not been applied yet in order. For instance, in our sample, if you start the application pointing to an empty database, it will apply V1__InitialDatabase.sql first and then V2__AddMatches.sql.

Flyway also uses flyway_schema_history table to control the concurrency while applying migrations. If you have more than one instance of your application, all of them will try to perform the same procedure:

1. The first application instance checks if the expected version is the same than the version deployed by looking at the flyway_schema_history table.

2. If the version deployed is the expected, they will continue the application normally.

3. If the version is different, it will lock the flyway_schema_history table, and will apply the migration.

4. The rest of application instances wait until the `flyway_schema_history` table is released.

5. When the migration finishes, the first application instance will update the version in `flyway_schema_history` table and will release it.

6. Then, the rest of application instances will check the version as in step 1. As it is already deployed, they will continue normally without applying the migration again.

Another validation performed by flyway is checking if a migration file has been modified. The way it does it is by generating a checksum of the content and saving in the `flyway_schema_history`. A checksum is a kind of signature generated from the content that can be used to verify that content has not been modified. The purpose of this validation is to ensure the consistency and repeatability of the process.

> **Important**
> Once applied a migration, don't modify the script file. If you need to fix a migration, create a new one performing the fix.

Keep in mind that large migrations, for instance those requiring data transformation, may produce locks on the database and potential downtimes in your application, as the application won't complete the initialization until the migration completes.

There's more...

Flyway provides a powerful mechanism to perform consistent migrations and keep your code in sync with the database schema. It provides robust mechanisms for versioning and rollback/undo operations as well as repeatable migrations.

If your application has complex requirements related to other components and not only your application, Flyway provides a mechanism named *Callback* to invoke additional actions related to the migrations, such as recompiling stored procedures, recalculating materialized view or flushing a cache, etc. If you have that kind of requirements, I recommend you to take a look at this documentation: `https://documentation.red-gate.com/flyway/flyway-cli-and-api/concepts/callback-concept`.

One drawback of using Flyway is that it may slow down the application boot process, even if there is no migration to apply. For that reason, Flyway also provides standalone tools to manage migrations, such as a Desktop UI and a command line tool. These tools aid in the process of allowing the migrations and related actions without adding any dependency to your project and performing the migration process independently.

See also

In this exercise, I focused on Flyway as a tool to manage database versioning, but Spring Boot also offers support for Liquibase. Just as Flyway, Liquibase can perform migrations during application startup and using independent tools such as the CLI. It has free and paid versions. I recommend you evaluate both tools and use the one that better meets your requirements.

Using JPQL

JPQL stands for **Java Persistence Query Language**. It is a platform-independent query language that is used to query and manipulate data stored in relational databases using the **Java Persistence API (JPA)**.

JPQL is similar in syntax to SQL, but it operates at the object level, allowing developers to write queries in terms of Java objects and their relationships rather than database tables and columns. This makes JPQL a more natural choice for developers working with Java-based applications and object-relational mapping frameworks like Hibernate.

Some key features and concepts of JPQL include:

- **Entity Classes**: JPQL queries are written against Java entity classes, which are Java objects that represent database tables.

- **Object-Oriented Queries**: JPQL allows you to query and manipulate data in an object-oriented way, using the names of Java classes and their attributes.

- **Relationships**: JPQL supports querying data based on the relationships between entities, such as one-to-one, one-to-many, and many-to-many associations.

- **Portability**: JPQL queries are written in a way that is independent of the underlying database system, making it possible to switch databases without changing the queries.

- **Type Safety**: JPQL queries are type-checked at compile-time, reducing the risk of runtime errors.

JPQL is a powerful tool for working with data in Java-based applications. It allows developers to express database queries in a way that is more aligned with the object-oriented nature of Java programming.

Getting ready

For this recipe, we don't need additional tools compared to previous recipes. As a starting point of this exercise, we will use the completed version of previous recipe. If you didn't complete it, you can find it in the book's repository at `https://github.com/PacktPublishing/Spring-Boot-3.0-Cookbook/`.

How to do it...

In this recipe, we will enhance the Repositories created in the previous recipe with some advanced queries using JPQL. We will add two more entities, `AlbumEntity` and `CardEntity` to emulate a card trading game. The data model will look like this:

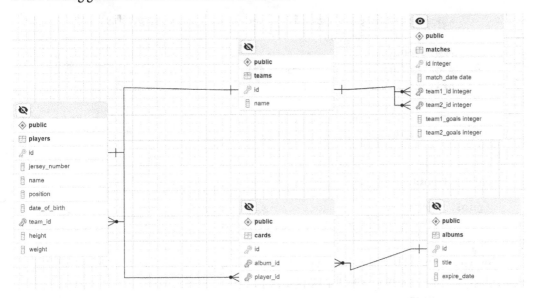

Figure 5.5: PostgreSQL Data model

Let's begin:

1. Add the new Entities and a new repository. We will need to create a new migration.

   - AlbumEntity:

```
@Table(name = "albums")
@Entity
public class AlbumEntity {
    @Id
    @GeneratedValue(strategy = GenerationType.IDENTITY)
    private Integer id;
    private String title;
    private LocalDate expireDate;
    @OneToMany
    private List<CardEntity> cards;
}
```

- CardsEntity:

```
@Table(name = "cards")
@Entity
public class CardEntity {
    @Id
    @GeneratedValue(strategy = GenerationType.IDENTITY)
    private Integer id;
    @ManyToOne
    @JoinColumn(name = "album_id")
    private AlbumEntity album;
    @ManyToOne
    @JoinColumn(name = "player_id")
    private PlayerEntity player;
}
```

- AlbumRepository:

```
public interface AlbumRepository extends
JpaRepository<AlbumEntity, Integer> {
}
```

Now create a new Flyway migration to create the tables. To do that, create a file named V3__AddAlbums.sql and create the tables.

```
CREATE TABLE albums (
    id SERIAL PRIMARY KEY,
    title VARCHAR(255),
    expire_date DATE
);
CREATE TABLE cards (
    id SERIAL PRIMARY KEY,
    album_id INTEGER REFERENCES albums(id),
    player_id INTEGER REFERENCES players(id)
);
```

This script is available in the book repository, including some sample data.

2. In our trading card game, the CardEntitity entity represents the cards that the user has. We will create a method to get the players we have of a certain team. To do that, in the AlbumRepository add the following method:

```
@Query("SELECT p FROM PlayerEntity p JOIN p.cards c WHERE
c.album.id = :id AND p.team.id = :teamId")
public List<PlayerEntity> findByIdAndTeam(Integer id, Integer
teamId);
```

3. We want to know what players we don't have yet. To find out, add the following method in the AlbumsRepository:

```
@Query("SELECT p FROM PlayerEntity p WHERE p NOT IN (SELECT
c.player FROM CardEntity c WHERE c.album.id=:id)")
public List<PlayerEntity> findByIdMissingPlayers(Integer id);
```

4. Let's find which are the players of a certain match, from both teams. In the MatchRepository add the following method:

```
@Query("SELECT p1 FROM MatchEntity m JOIN m.team1 t1 JOIN
t1.players p1 WHERE m.id = ?1 UNION SELECT p2 FROM MatchEntity m
JOIN m.team2 t2 JOIN t2.players p2 WHERE m.id = ?1")
public List<PlayerEntity> findPlayersByMatchId(Integer matchId);
```

5. Get a team and its players. To do that, in the TeamRepository add the following method:

```
@Query("SELECT t FROM TeamEntity t JOIN FETCH t.players WHERE
t.id = ?1")
public Optional<TeamEntity> findByIdWithPlayers(Integer id);
```

This method can be used now in FootballService to get the teams. If you remember from the recipe *Connect your application to Postgresql*, we implemented a mechanism to avoid the Open Session In View antipattern by adding a @Transactional annotation in the method getTeam. With this new TeamRepository method, it retrieves both the team and its players in the same session, then it is not necessary @Transactional.

```
public Team getTeam(Integer id) {
    TeamEntity team = teamRepository.findByIdWithPlayers(id).
orElse(null);
    if (team == null) {
        return null;
    } else {
    return new Team(team.getId(),
                    team.getName(),
                    team.getPlayers()
                        .stream()
                        .map(player -> new Player(player.
getName(), player.getJerseyNumber(), player.getPosition(),
                                player.getDateOfBirth())))
                        .toList());
    }
}
```

6. Find a list of players. Modify PlayerRepository by adding the following method:

```
@Query("SELECT p FROM PlayerEntity p WHERE p.id IN (?1)")
List<PlayerEntity> findListOfPlayers(List<Integer> players);
```

This method could be also implemented just by using naming convention with no need for @ Query annotation.

```
List<PlayerEntity> findByIdInList(List<Integer> players);
```

7. Find a list of players whose names contain a certain string. Modify the `PlayerRepository` and add the following method:

```
List<PlayerEntity> findByNameLike(String name);
```

8. Find a list of players whose name starts with a string. Modify the `PlayerRepository` and add the following method:

```
List<PlayerEntity> findByNameStartingWith(String name);
```

9. Sort the players of a team in ascending order. Modify the `PlayerRepository` and add the following repository.

```
List<PlayerEntity> findByTeamId(Integer teamId, Sort sort);
```

You can use this method as follows:

```
playerRepository.findByTeamId(id, Sort.by("name").ascending())
```

You can decide how to sort the results.

10. We have the option to paginate the results. This means that if the result set is extensive, we can divide it into pages and retrieve. The `JpaRepository` already provides method overloads to page the results. For instance, the method `findAll` can receive a pageable parameter to control how the results should be paged.

```
Page<PlayerEntity> page = playerRepository.findAll(Pageable.
ofSize(size).withPage(pageNumber));
```

You can add this as parameter in any of the methods using a custom query. For instance, we can create the following method in `AlbumsRepository`:

```
@Query("SELECT p FROM PlayerEntity p JOIN p.cards c WHERE
c.album.id = :id")
public List<PlayerEntity> findByIdPlayers(Integer id, Pageable
page);
```

11. We can also use JPQL to return aggregated results. For instance, let's create a query to get the number of players in a given position, per team.

```
@Query("SELECT p.team.name as name, count(p.id) as playersCount
FROM PlayerEntity p WHERE p.position = ?1 GROUP BY p.team ORDER
BY playersCount DESC")
public List<TeamPlayers> getNumberOfPlayersByPosition(String
position);
```

As you can see, the result is not an entity, but just the name of the team and the number of players in the position. To return this result we used a custom result, implemented as an Interface.

```
public interface TeamPlayers {
    String getName();
    Integer getPlayersCount();
}
```

The interface should have getter methods matching the projected query result.

12. Create a RESTful controller and a service to use the methods generated. In the book's GitHub repository there is a RESTful API using the repositories created in this recipe.

    In the repository you can also find a script to call the RESTful API methods created in this recipe.

How it works...

When an application using Spring Data JPA starts, it performs several important actions to make it work.

- Spring Application Context initialization. It sets up the environment for managing Spring components, including the repositories.

- Spring Boot scans component and detect repositories. It checks classes annotated with @ Repository and interfaces extending JpaRepository.

- For each repository interface, Spring Data JPA generates a concrete implementation during runtime. In our scenario, it means that it takes each method defined in the repositories and generates the specific queries. At this step it validates if it can generate the implementation by using the naming convention, or by using the @Query annotation. At this step it also validates the queries, so if we write an invalid query or it is not capable of generating the implementation from the naming convention, it will fail.

- After generating the implementation, it registers them as beans in the application context and are now available for the rest of components of our application.

One important advantage of JPA and JPQL is that the queries reference the Entities we defined in our code, so it can detect a query/entity mapping mismatch early. This cannot be achieved when using native queries.

Another advantage is that it abstracts the underlaying database. As a developer, this is an interesting feature, as it makes the onboarding to a new database faster.

There's more...

You can activate the SQL logging by using spring.jpa.show-sql configuration variable. It is interesting to check and debug the native queries being generated. Keep in mind that it can slow down your application and generate large logs. I recommend using this setting only in development.

Using Native Queries

JPQL is a very powerful mechanism to access relational databases with an abstraction of the underlying database. Native Queries refers to executing SQL statements directly against the database. Depending on your requirements you might consider using Native queries, for instances like:

- Executing complex SQL operations that are not easily expressible in JPQL, for instance, queries that involve subqueries or database-specific functions.

- When you need to fine-tune the performance of a query, leveraging database-specific optimizations, indexes, and hints to improve the query execution time.

- Database-specific features, for instance, the databases that can manage JSON structures, may have different ways of doing it.

- Bulk Operations: Native queries are often more efficient for executing bulk insert, update, or delete operations on a large number of records, as they bypass the overhead of entity management and caching that comes with JPQL.

Keep in mind that using Native Queries comes with some trade-offs compared to JPQL.

- **Type Safety**: As we saw in the previous recipe, JPQL provides type safety, which means that query results are returned as strongly typed objects and queries are validated during application startup. With native queries, you typically work with untyped result sets, which can introduce runtime errors if not handled properly and that will come up just when the native query is used.

  Native queries can be harder to maintain and refactor because they involve SQL strings embedded in your Java code. JPQL queries are more self-contained and easier to manage

- **Portability**: Native queries are not portable across different database systems. If your application needs to support multiple databases, you may need to write database-specific queries for each one, and you will need to learn the specific differences of each database SQL dialect.

In this recipe, we'll introduce a new feature, match timelines, to our Football application. The timeline are all the events that happen during a football match; as we don't want to constrain the content that can be managed as an event, we will save part of the information as JSON. PostgreSQL has excellent support for JSON, but for most of the scenarios, it is necessary to write native queries.

Getting ready

For this recipe, we don't need additional tools compared to previous recipes. As starting point of this exercise, we will use the completed version of previous recipe. If you didn't complete it, you can find it in the book's repository at `https://github.com/PacktPublishing/Spring-Boot-3.0-Cookbook/`. I also prepared some scripts to create some sample data for the database. It will be explained as part of the recipe steps.

How to do it...

In this recipe, we will create a new table managing the match events. That table will have a column containing JSON documents.

1. To create the new table, we will create a new Flyway migration. In folder `src/main/resources/db/migration` create a file named `V4__AddMatchEvents.sql` with the following content.

    ```
    CREATE TABLE match_events (
        id BIGSERIAL PRIMARY KEY,
        match_id INTEGER NOT NULL,
        event_time TIMESTAMP NOT NULL,
        details JSONB,
        FOREIGN KEY (match_id) REFERENCES matches (id)
    );
    CREATE PROCEDURE FIND_PLAYERS_WITH_MORE_THAN_N_MATCHES(IN num_
    matches INT, OUT count_out INT)
    LANGUAGE plpgsql
    AS $$
    BEGIN
        WITH PLAYERS_WITH_MATCHES AS
            (SELECT p.id, count(m.id) AS match_count FROM players p,
    matches m WHERE p.team_id = m.team1_id OR p.team_id = m.team2_id
            GROUP BY p.id HAVING count(m.id) > num_matches)
        SELECT COUNT(1) INTO count_out FROM PLAYERS_WITH_MATCHES;
    END;
    $$;
    ```

    Additionally, I prepared another migration that you can find in `https://github.com/PacktPublishing/Spring-Boot-3.0-Cookbook/` named `V4_1__CreateSampleEvents.sql`. This migration inserts events in the `match_events` table, so you will be able to play around it. Here you can see an example of match event detail:

    ```
    {
        "type": 24,
        "description": "Throw In",
        "players": [
          467653,
          338971
        ],
        "mediaFiles": [
          "/media/93050144.mp4",
          "/media/6013333.mp4",
    ```

```
        "/media/56559214.mp4"
    ]
}
```

2. Create a new entity to manage this table. Create a class named `MatchEventEntity`:

```
@Table(name = "match_events")
@Entity
public class MatchEventEntity {
    @Id
    @GeneratedValue(strategy = GenerationType.IDENTITY)
    private Long id;

    @Column(name = "event_time")
    private LocalDateTime time;

    @JdbcTypeCode(SqlTypes.JSON)
    private MatchEventDetails details;

    @ManyToOne(fetch = FetchType.LAZY)
    @JoinColumn(name = "match_id", nullable = false)
    private MatchEntity match;
}
```

3. We will map the JSON content in another class named `MatchEventDetails`. You can use other more flexible data structures, such as a Map.

```
public class MatchEventDetails {
    private Integer type;
    private String description;
    private List<Integer> players;
    private List<String> mediaFiles;
}
```

4. Create a new `JpaRepository` and name it `MatchEventRepository`:

```
public interface MatchEventRepository extends
JpaRepository<MatchEventEntity, Long> {
}
```

5. We will create a new method in the repository to retrieve all events in a match of a given type. The type of event is just an attribute of the JSON content. To perform this query, we need to use PostgreSQL specific syntax to query using the JSON content. To use Native Queries, we just need to specify in @Query annotation the attribute nativeQuery = true.

```
@Query(nativeQuery = true, value = "SELECT me.* FROM match_
events me  WHERE me.match_id = ?1 AND CAST(me.details -> 'type'
as INT) = ?2")
public List<MatchEventEntity>
findByIdIncludeEventsOfType(Integer matchId, Integer eventType);
```

6. We will implement a repository method to retrieve the events associated with a particular football match in which a specified player participated.

```
@Query(nativeQuery = true,
   value = "SELECT me.id, me.match_id, me.event_time, " +
me.details FROM match_events me CROSS JOIN LATERAL " + jsonb_
array_elements(me.details->'players') AS player_id " +
"WHERE me.match_id = ?1 AND CAST(player_id as INT) = ?2")
List<MatchEventEntity> findByMatchIdAndPlayer(Integer matchId,
Integer playerId);
```

7. In PlayerRepository we will create a method a new method to map the stored procedure.

```
@Procedure("FIND_PLAYERS_WITH_MORE_THAN_N_MATCHES")
int getTotalPlayersWithMoreThanNMatches(int num_matches);
```

Now you can create a service and a controller using these repositories. In the GitHub repository, I extended the existing controller to call the new repository methods. You can also find a script calling the new controller methods.

How it works...

In this example, we used JSON because it is very useful to store data that can be flexible, extensible and it is not required to be as structured as a table, with known columns and types. PostgreSQL has a good support for JSON, however JPQL support for this kind of scenario is more limited. That is why, you need to use Native Queries.

> **Note**
>
> Even if PostgreSQL has very good support for JSON, it is not as optimized as it is for regular columns. If there is information in the document that is frequently used, it is better to move it to regular columns. PostgreSQL supports indexes over JSON properties, you will need to evaluate which is the best approach for your specific scenario.

We used **Plain Old Java Objects** (**POJO**) to represent the JSON column. `MatchEventDetail` is just a class, not an Entity. In any case, this also requires knowing the schema of the JSON data in advance to avoid serialization errors. If you need a more flexible approach, you can use just a Map or a String to map that column.

JSON support is just an example of a native feature in this case of PostgreSQL, but there are other scenarios that you might require using native queries. For instance, complex queries that are difficult or not possible to express using JPQL, such as subqueries and bulk operations.

When a Native Query is executed, the SQL command itself is not checked by Spring Data JPA nor Hibernate, however the resulting execution should be mapped to the resulting Entities. Keep that in mind when you write the SQL statements. For instance, if in this recipe, we wrote:

```
SELECT me.id, me.match_id, me.event_time, me.details FROM match_events
me CROSS JOIN LATERAL  jsonb_array_elements(me.details->'players') AS
player_id
WHERE me.match_id = ?1 AND CAST(player_id as INT) = ?2)
```

That should match a `List<MatchEventEntity>`. According to our definition of `MatchEventEntity`, it expects that the result of the query contains the columns *id*, *event_time*, *details* and *match_id*. So be aware of this fact when using alias in your queries. As an example, look at the following query, it will result in an error in runtime:

```
SELECT me.id as event_id, me.match_id, me.event_time, me.details FROM
match_events.
```

In this recipe, we also used a stored procedure. In this case, the mapping is simpler and more direct. Just by using `@Procedure` annotation, Spring Data JPA and Hibernate can invoke the stored procedure. As it happens with Native Queries, you are responsible to make sure that the incoming parameters and results match with the method invocation. If you change any of them, it can cause errors in runtime.

There's more...

An important difference seen in Native queries compared to JPQL queries is that the queries cannot be validated against your entities, hence you need to be careful as it can fail at runtime. I recommend checking all queries first in tools such as `PgAdmin` for PostgreSQL, or a similar tool for the database you choose. I also recommend preparing a good set of tests using the native queries. In the book's GitHub repository I created some tests to validate the queries used in this recipe.

See also

If your solution is more dependent on JSON documents, with schema flexibility, rather than well-defined schemas with complex relationships and transactional integrity needs that PostgreSQL and other relational databases can offer, then you may want to consider other database technologies, such as Document databases. There are many solutions in the market, like MongoDB, Azure CosmosDB, AWS DocumentDb. We will cover MongoDB in the following chapter.

Updating Operations

In previous recipes, we just performed queries against the database. In this recipe, we will use Spring Data JPA to modify the data of our database.

We will continue with our football sample. In this recipe, we will create operations to manage the trading card albums. A user may have albums, an album has cards, and the card references a player. To complete an album, the user needs to have cards with all players. So, they need to know what cards are missing. Users can buy albums and cards, but they cannot choose the cards. They can have repeated players. Users can trade cards. So, they can exchange all unused cards with another user.

Getting ready

For this recipe, we don't need additional tools compared to previous recipes. As a starting point of this exercise, we will use the completed version of the previous recipe. If you didn't complete it, you can find it in the book's repository at `https://github.com/PacktPublishing/Spring-Boot-3.0-Cookbook/`.

How to do it...

To complete this recipe, we will add a new Entity to manage the users. We will modify Albums and Cards as they have an owner now. Later we will create some operations that involve data modification to manage the cards and trading operations.

1. We will first create a new database migration using Flyway. To do that, create a file named `V5__AddUsers.sql` in `src/main/resources/db/migration`.

    Then, we will create a table for users, and we'll update the cards and albums to reference the user.

    ```sql
    CREATE TABLE users (
        id SERIAL PRIMARY KEY,
        username VARCHAR(255)
    );
    ALTER TABLE albums ADD COLUMN owner_id INTEGER REFERENCES users(id);
    ALTER TABLE cards ADD COLUMN owner_id INTEGER REFERENCES users(id);
    ALTER TABLE cards ADD CONSTRAINT cards_album_player_key UNIQUE (album_id, player_id);
    ```

    Note the constraint in the table cards to avoid repeating players in the same album.

2. Add a new entity named `UserEntity` to map with the new table:

    ```java
    @Table(name = "users")
    @Entity
    public class UserEntity {
    ```

```
@Id
@GeneratedValue(strategy = GenerationType.IDENTITY)
private Integer id;

private String username;

@OneToMany(mappedBy = "owner")
private List<CardEntity> ownedCards;

@OneToMany(mappedBy = "owner")
private Set<AlbumEntity> ownedAlbums;
}
```

3. Modify CardEntity and AlbumEntity to reference an owner user.

```
@ManyToOne
@JoinColumn(name = "owner_id")
private UserEntity owner;
```

4. CardEntity can be modified to reflect the unique constraint that a player can be only once in an album:

```
@Table(name = "cards", uniqueConstraints = { @
UniqueConstraint(columnNames = { "album_id", "player_id" }) })
@Entity
public class CardEntity {
}
```

5. Let's start managing the data. We will start with just using the methods already provided by JpaRepository.

   Let's create a UserRepository:

```
public interface UserRepository extends
JpaRepository<UserEntity, Integer> {
}
```

JpaRepository provides a method named save. This method creates or updates the entity provided. We can use it in this way:

```
private UserRepository usersRepository;
public User createUser(String name) {
    UserEntity user = new UserEntity();
    user.setUsername(name);
    user = usersRepository.save(user);
    return new User(user.getId(), user.getUsername());
}
```

In the same way, we can create an album that references a user:

```
public Album buyAlbum(Integer userId, String title) {
    AlbumEntity album = new AlbumEntity();
    album.setTitle(title);
    album.setExpireDate(LocalDate.now().plusYears(1));
    album.setOwner(usersRepository.findById(userId).
orElseThrow());
    album = albumsRepository.save(album);
    return new Album(album.getId(), album.getTitle(), album.
getOwner().getId());
}
```

We can also save multiple entities at the same time by calling the method `saveAll`. As an example, let's define a method to buy cards:

```
public List<Card> buyCards(Integer userId, Integer count) {
    Random rnd = new Random();
    List<PlayerEntity> players = getAvailablePlayers();
    UserEntity owner = usersRepository.findById(userId).
orElseThrow();
    List<CardEntity> cards = Stream.generate(() -> {
        CardEntity card = new CardEntity();
        card.setOwner(owner);
        card.setPlayer(players.get(rnd.nextInt(players.size())));
        return card;
    }).limit(count).toList();
    return cardsRepository.saveAll(cards)
                .stream()
                .map(card -> new Card(card.getId(), card.
getOwner().getId(), Optional.empty(),
                        new Player(card.getPlayer().getName(),
card.getPlayer().getJerseyNumber(),
                            card.getPlayer().getPosition(),
card.getPlayer().getDateOfBirth())))
                .collect(Collectors.toList());
}
```

Our users will buy batches of cards. We need to generate the cards; we will generate the cards selecting a random player for each card, then we'll save them in a single `saveAll` operation.

6. Once we have the cards, we want to use them in our albums. Using them constitutes assigning them to an album. If we want to use just the method provided by JpaRepository, we should perform the following steps:

I. Get all the available cards, that is the ones that have not been assigned to an album.

II. Get all missing players. That is all players that are not in the cards assigned to an album.

III.   Take all available cards that are in the missing players. These are the cards to be assigned to albums.

IV.   Verify that you only use a player in an album once.

V.   Save the cards.

All these steps involve requests to the database.

Or we can obtain the same result with just one request to the database by using an UPDATE SQL command:

```
UPDATE cards
SET album_id = r.album_id
FROM
(SELECT available.album_id, (SELECT c2.id from cards c2 where
c2.owner_id=?1 AND c2.player_id = available.player_id AND
c2.album_id IS NULL LIMIT 1) as card_id
FROM
(SELECT DISTINCT a.id as album_id, c.player_id FROM albums a
CROSS JOIN cards c WHERE a.owner_id=?1 AND c.owner_id=?1 AND
c.album_id IS NULL AND c.player_id NOT IN (SELECT uc.player_id
from cards uc WHERE uc.album_id = a.id)) available) as r
WHERE cards.id = r.card_id
```

We can use this command in our `CardRepository`:

```
@Modifying
@Query(nativeQuery = true, value = "UPDATE cards " +
  "SET album_id = r.album_id  " + //
  "FROM " + //
  "(SELECT available.album_id, (SELECT c2.id from cards c2
where c2.owner_id=?1 AND c2.player_id = available.player_id AND
c2.album_id IS NULL LIMIT 1) as card_id " + //
  "FROM " + //
  "(SELECT DISTINCT a.id as album_id, c.player_id FROM albums
a CROSS JOIN cards c WHERE a.owner_id=?1 AND c.owner_id=?1 AND
c.album_id IS NULL AND c.player_id NOT IN (SELECT uc.player_id
from cards uc WHERE uc.album_id = a.id)) available) as r " +
  "WHERE cards.id = r.card_id " +
  "RETURNING cards.*")
  List<CardEntity> assignCardsToUserAlbums(Integer userId);
```

Remember to include @Modifying annotation. As this is a PostgreSQL command, it requires `nativeQuery=true` in the @Query annotation.

7. We can transfer a card to another user. If the card was used in an album, it should be unlinked. This can be done in different ways, we will implement the same using a JPQL Query:

```
@Modifying
@Query(value = "UPDATE CardEntity " +
" SET album = null, " +
" owner= (SELECT u FROM UserEntity u WHERE u.id=?2) " +
"WHERE id = ?1 ")
Integer transferCard(Integer cardId, Integer userId);
```

In this case, we need to ensure that this method is executed in the context of a transaction. We can do it decorating the calling method with a `@Transactional` annotation:

```
@Transactional
public Optional<Card> transferCard(Integer cardId, Integer userId) {
    Integer count = cardsRepository.transferCard(cardId, userId);
    if (count == 0) {
        return Optional.empty();
    } else {

        ...

    }
}
```

8. Next, we'll learn how to exchange cards from user to another. Again, we can do it in our business logic using the methods provided by `JpaRepository` by performing the following actions:

I. Get the available cards from one user, these are the ones not assigned to an album usually because they are repeated players.

II. Get the missing players on the albums of the other user.

III. Change the owner of the cards of the first user that are in the list of the missing players of the other user.

Or we can do it in a single SQL UPDATE statement:

```
@Modifying
@Query(nativeQuery = true, value = "UPDATE cards " +
"SET owner_id=?2 " +
" FROM (select c1.id from cards c1 where c1.owner_id=?1 and
c1.album_id IS NULL AND c1.player_id IN (select p2.id from
players p2 where p2.id NOT IN (SELECT c2.player_id FROM cards c2
WHERE c2.owner_id=?2)) LIMIT ?3) cards_from_user1_for_user2 " +
"WHERE cards.id = cards_from_user1_for_user2.id " +
"RETURNING cards.*")
List<CardEntity> tradeCardsBetweenUsers(Integer userId1, Integer
userId2, Integer count);
```

I created a service class and a dedicated RESTful controller to perform all the operations above. The code, including a script to call the RESTful controller, is in the GitHub repository.

How it works...

In JPA, there is the concept of *Persistence Context* or just *Persistence Context*. The Context is mostly managed by *EntityManager*, which is responsible for managing the lifecycle of the JPA entities. It covers the following aspects:

- **Entity Management**: The persistence context is responsible for managing the entities. When you retrieve data from the database using Spring Data JPA, the resulting entities are managed by the persistence context. This means that changes to these entities are tracked, and you can use the persistence context to synchronize these changes with the database.

- **Identity Management**: The persistence context ensures that there is a single in-memory representation of an entity for a given database row. If you load the same entity multiple times, you will get the same Java object instance, ensuring consistency and avoiding duplicate data.

- **Automatic Dirty Checking**: The persistence context automatically tracks changes made to managed entities. When you modify an entity's state, these changes are detected, and the associated database records are updated when the persistence context is flushed. You can flush explicitly using the `flush` method of `JpaRepository`, or implicitly, for instance at the end of a transaction.

- **Caching**: The persistence context provides a first-level cache. It stores managed entities in memory, which can improve application performance by reducing the number of database queries required for entity retrieval during a transaction.

- **Lazy Loading**: The persistence context can enable lazy loading of related entities. When you access a property representing a related entity, Spring Data JPA can automatically fetch that related entity from the database, if it's not already in the persistence context.

- **Transaction Synchronization**: The persistence context is typically bound to the scope of a transaction. This ensures that changes to entities are persisted to the database when the transaction is committed. If the transaction is rolled back, the changes are discarded.

In Spring Data JPA, the `EntityManager` is the central component for managing the persistence context. It provides methods for persisting, retrieving, and managing entities.

In addition to the methods already provided by the `JpaRepository` like `save` and `saveAll`, you can use `@Modifying` annotation. In Spring Data JPA, the `@Modifying` annotation is used to indicate that a method in a Spring Data JPA repository interface is a modifying query method. Modifying query methods are used to perform data modification operations like INSERT, UPDATE, or DELETE in the database.

When you mark a method in a Spring Data JPA repository with the @Modifying annotation, it changes the behavior of that method in the following ways:

- **Non-SELECT Query Execution**: The @Modifying annotation indicates that the method is intended to execute a non-select query, such as an UPDATE or DELETE statement.

- **No Automatic Result Type Inference**: With normal query methods in Spring Data JPA, the framework automatically infers the return type based on the method name. When you use the @Modifying annotation, the return type is not inferred automatically. Instead, you should explicitly specify the return type. Typically, the return type is int or void. For example, in the example above, transferCard returns an Integer. That number represents the number of rows affected. With Native Queries it is possible to return data, as shown in method tradeCardsBetweenUsers. PostgreSQL can return the rows impacted using the keyword RETURNING in the UPDATE command. This behavior can change depending on the Database engine.

- **Flush Behavior**: By default, Spring Data JPA does not automatically flush the persistence context to the database after executing a query method. However, when you use @Modifying, it will trigger a flush of the persistence context to synchronize the changes with the database.

- **Transaction Requirement**: Modifying query methods should be executed within a transaction context. If the method is called without an active transaction, it will typically result in an exception when using JPQL queries. For Native Queries, this behavior does not apply. We will cover in more detail transaction management in another recipe of this chapter.

In this recipe, we used JPQL queries and native queries. As mentioned in previous recipes, JPQL has the primary advantage of using your Entities, being able to make a type-safety check, and abstracting the complexities of accessing the underlying database.

Native Queries can be necessary when you need to fine tune your queries and optimize the access to the database. Taking the example of assignCardsToUserAlbums, the same operation using just JPQL and business logic in your Java application will require several calls to the database, transferring data from the database and to the database. This communication overhead is not a negligible cost for large-scale applications. In the implementation of assignCardsToUserAlbums, it is just one single call to PostgreSQL that performs all the updates and returns just the cards updated to be returned to the caller component.

See also

Check the *Using Dynamic Queries* and *Using Transactions* recipes to deeper dive into EntityManager and transactions management in Spring Data JPA.

Using Dynamic Queries

In Spring Data JPA, a dynamic query refers to a type of query that is constructed at runtime based on various conditions or parameters. Dynamic queries are particularly useful when you need to build and execute queries that can vary based on user input or changing criteria. Instead of writing a static query with fixed criteria, you create a query that adapts to different scenarios.

Dynamic queries can be constructed using the Criteria API, or by creating the query statement dynamically using JPQL or Native SQL. The Criteria API provides a programmatic and type-safe way to define queries to the database.

In previous recipes, we used the naming convention of the `JpaRepositories` when creating the repository methods. Spring Data JPA generates the queries dynamically using the same mechanism we will explain in this recipe.

In this recipe, we will implement the following functionalities:

- Search players using different criteria, for instance by name, height, or weight.

- Search match events in a time range.

- Delete match events in a time range.

- Search the missing players that a user of the card trading game does not have yet.

Getting ready

For this recipe, we don't need additional tools compared to previous recipes. As the starting point of this exercise, we will use the completed version of *Updating Operations* recipe. If you didn't complete it, you can find it in the book repository at `https://github.com/PacktPublishing/Spring-Boot-3.0-Cookbook/`.

How to do it...

For this recipe, we will create a new service component to perform all dynamic queries.

1. To do that, first create a new `Service` named `DynamicQueriesService`. The service requires an `EntityManager`. For that reason, we need to declare a parameter in the constructor to ensure that the IoC container injects it.

    ```
    @Service
    public class DynamicQueriesService {

    }
    ```

2. In this service, we can create a method to search players using different criteria. Each criteria are optional, so we need to construct the query dynamically depending on the parameters provided. We will use the `CriteriaBuilder` class for that purpose.

```java
Public List<PlayerEntity> searchTeamPlayers(Integer teamId,
Optional<String> name, Optional<Integer> minHeight,
            Optional<Integer> maxHeight,
            Optional<Integer> minWeight, Optional<Integer>
maxWeight) {
    CriteriaBuilder cb = em.getCriteriaBuilder();
    CriteriaQuery<PlayerEntity> cq =
cb.createQuery(PlayerEntity.class);
    Root<PlayerEntity> player = cq.from(PlayerEntity.class);
    List<Predicate> predicates = new ArrayList<>();
    predicates.add(cb.equal(player.get("team").get("id"),
teamId));
    if (name.isPresent()) {
        predicates.add(cb.like(player.get("name"), name.get()));
    }
    if (minHeight.isPresent()) {
        predicates.add(cb.ge(player.get("height"), minHeight.
get()));
    }
    if (maxHeight.isPresent()) {
        predicates.add(cb.le(player.get("height"), maxHeight.
get()));
    }
    if (minWeight.isPresent()) {
        predicates.add(cb.ge(player.get("weight"), minWeight.
get()));
    }
    if (maxWeight.isPresent()) {
        predicates.add(cb.le(player.get("weight"), maxWeight.
get()));
    }
    cq.where(predicates.toArray(new Predicate[0]));
    TypedQuery<PlayerEntity> query = em.createQuery(cq);
    return query.getResultList();
}
```

In this example, we used the criteria query as a parameter for the `EntityManager` method `createQuery`, then we used the query to retrieve the results.

3. Let's implement another example, this time using JPQL statements. We will search events of a match in a time range:

```
public List<MatchEventEntity> searchMatchEventsRange(Integer
matchId, Optional<LocalDateTime> minTime,
Optional<LocalDateTime> maxTime) {
    String command = "SELECT e FROM MatchEventEntity e WHERE
e.match.id=:matchId ";
    if (minTime.isPresent() && maxTime.isPresent()) {
        command += " AND e.time BETWEEN :minTime AND :maxTime";
    } else if (minTime.isPresent()) {
        command += " AND e.time >= :minTime";
    } else if (maxTime.isPresent()) {
        command += " AND e.time <= :maxTime";
    }
    TypedQuery<MatchEventEntity> query = em.createQuery(command,
MatchEventEntity.class);
    query.setParameter("matchId", matchId);
    if (minTime.isPresent()) {
        query.setParameter("minTime", minTime.get());
    }
    if (maxTime.isPresent()) {
        query.setParameter("maxTime", maxTime.get());
    }
    return query.getResultList();
}
```

Now the query is created using a String that is passed again to the `createQuery` method of the `EntityManager`.

As you can see, the command contains named parameters that must be passed to the query.

4. We can use Native SQL commands as well. Let's search for the players that a user doesn't have yet for his or her album.

```
public List<PlayerEntity> searchUserMissingPlayers(Integer
userId) {
    Query query = em.createNativeQuery(
    "SELECT p1.* FROM players p1 WHERE p1.id NOT IN (SELECT
c1.player_id FROM cards c1 WHERE c1.owner_id=?1)",
    PlayerEntity.class);
    query.setParameter(1, userId);
    return query.getResultList();
}
```

To execute the native query, we now pass the String containing the native SQL command to the `createNativeQuery` method.

5. Now we will create a method to perform a delete operation. We will delete the events of a match in a certain time range. We will use JPQL to perform this functionality.

```
public void deleteEventRange(Integer matchId, LocalDateTime
start, LocalDateTime end) {
    em.getTransaction().begin();
    Query query = em.createQuery("DELETE FROM MatchEventEntity e
WHERE e.match.id=:matchId AND e.time BETWEEN :start AND :end");
    query.setParameter("matchId", matchId);
    query.setParameter("start", start);
    query.setParameter("end", end);
    query.executeUpdate();
    em.getTransaction().commit();
}
```

To perform an update, we need to call the `executeUpdate` method. Note that this type of modifying operation requires an active transaction.

How it works...

As explained in the previous recipe, in Spring Data JPA, the `EntityManager` is the central component for managing the persistence context. It provides methods for persisting, retrieving, and managing entities.

When using JPQL, the `EntityManager` compiles the query, not only to validate the syntax but also to check the consistency with the Entities of our project, then translates the query to native SQL and binds the parameters. After executing the query, `EntityManager` maps the results into the managed entities. The resulting entities are tracked by the `EntityManager` for any change or further persistence operation.

If you have a query that will be executed multiple times, you can use a Named Query, as it is precompiled and cached for better performance. For that, you can call the `createNamedQuery` method.

For Native queries, it is a bit simpler as it doesn't compile nor validates the consistency, and it directly executes the query, mapping the results to the Entity specified. As discussed in previous recipes, it has advantages and trade-offs that you will need to evaluate depending on the needs of your application.

In the examples, we used Criteria Query and just a String containing the command. In general, I prefer Criteria Query because using it helps you avoid typos while building your query. In addition, Criteria Query is protected against SQL Injection attacks. If you build your query just concatenating Strings, be sure that you don't use parameters provided by the user directly as it will make your query vulnerable to SQL Injection attacks. When using parameters provided by the user, be sure that they are always passed as query parameters and never directly concatenated to the command string. See for example the method `searchMatchEventsRange`. In it, the parameters influence the SQL command generated, but they are always passed as query parameters.

There's more...

I created a controller to use the methods created in this recipe. You can find it in the book's GitHub repository. In that project I enabled swagger UI, so you can use it to test the methods.

See also

There is a lot of literature about SQL Injection. If you are not familiar with it, you can check the Wikipedia page: `https://en.wikipedia.org/wiki/SQL_injection`.

Using Transactions

Transactions play a crucial role when working with databases to ensure data consistency and integrity. In Spring Data JPA, you can manage transactions using the Spring Framework's transaction management capabilities.

In previous recipes, we implicitly used transactions, as some features require its usage; for instance, using @Modifiying annotation creates a transaction behind the scenes. In this recipe, we will learn more about transactions and how to use them in Spring Boot applications.

As an example of using transactions in Spring Boot applications, we will use them to manage the trading card operations between users. In a high-concurrency scenario, we want to ensure that users can exchange their cards with consistency and integrity.

Getting ready

For this recipe, we don't need additional tools compared to previous recipes. As a starting point of this exercise, you can use a project that I prepared with the previous recipes in this chapter. You can find it in the book's repository at `https://github.com/PacktPublishing/Spring-Boot-3.0-Cookbook/`.

How to do it...

To implement the card trading scenario, we will enhance the `AlbumsService` to manage the card exchange between users consistently.

1. Open the service `AlbumsService` and find the method `tradeAllCards`. The functionality we want to achieve with this method is the following:

    - Users can have repeated cards.

    - They can exchange cards between them.

    - They want to exchange the cards necessary to complete the albums.

- A trading operation involves the same number of cards from each user. If Sara gives three cards to Paul, Paul should give three cards to Sara in return.

To implement this functionality, first we need to know how many cards two users can exchange:

```
Integer potentialUser1ToUser2 = cardsRepository.
countMatchBetweenUsers(userId1, userId2);

Integer potentialUser2ToUser1 = cardsRepository.
countMatchBetweenUsers(userId2, userId1);

Integer count = Math.min(potentialUser1ToUser2,
potentialUser2ToUser1);
```

If both users have cards that can be exchanged, then they exchange a number of cards from one user to another, and then the same in the other direction. To avoid the same cards being exchanged in both directions, once a user receives the cards, they are used. Therefore these cards are not available for exchange.

```
ArrayList<CardEntity> result = new ArrayList<>(
                 cardsRepository.
tradeCardsBetweenUsers(userId1, userId2, count));

useAllCardAvailable(userId2);

result.addAll(cardsRepository.tradeCardsBetweenUsers(userId2,
userId1, count));

useAllCardAvailable(userId1);
```

2. We want to perform all actions described in step 1consistently, and in case of an error, we want users to have the same cards they had before starting the trading operation. For this, we only need to annotate the method with @Transactional. The full method should look like this:

```
@Transactional
public List<Card> tradeAllCards(Integer userId1, Integer
userId2) {
    Integer potentialUser1ToUser2 = cardsRepository.
countMatchBetweenUsers(userId1, userId2);
    Integer potentialUser2ToUser1 = cardsRepository.
countMatchBetweenUsers(userId2, userId1);
    Integer count = Math.min(potentialUser1ToUser2,
potentialUser2ToUser1);
    if (count > 0) {
    ArrayList<CardEntity> result = new ArrayList<>(
                 cardsRepository.
tradeCardsBetweenUsers(userId1, userId2, count));
    useAllCardAvailable(userId2);
    result.addAll(cardsRepository.
tradeCardsBetweenUsers(userId2, userId1, count));
    useAllCardAvailable(userId1);
}
```

There is some boilerplate code used to return data to be consumed by the RESTful API. It has been omitted for brevity, but you can find the full method in the GitHub repository.

If there is an exception during the execution of this method, the transaction will be automatically rolled-back, keeping the state as it was before starting the transaction. Let's modify this method to add some validations.

We will check that the number of cards traded are the same. For that, change the invocation to `tradeCardsBetweenUsers` as follows:

```
ArrayList<CardEntity> result1 = new ArrayList<>(
    cardsRepository.tradeCardsBetweenUsers(userId1, userId2,
count));
useAllCardAvailable(userId2);
ArrayList<CardEntity> result2 = new ArrayList<>(
    cardsRepository.tradeCardsBetweenUsers(userId2, userId1,
count));
useAllCardAvailable(userId1);
if (result1.size() != result2.size()) {
    throw new RuntimeException("Users have different number of
cards");
}
```

3. We can achieve a similar functionality by controlling the isolation of the transaction. You can do it by changing `@Transactional` annotation properties. If you set the transaction isolation level to `Serializable`, you ensure that the data used in your transaction is not read by any other transaction.

```
@Transactional(isolation = Isolation.SERIALIZABLE)
public List<Card> tradeAllCards(Integer userId1, Integer
userId2) {
    Integer potentialUser1ToUser2 = cardsRepository.
countMatchBetweenUsers(userId1, userId2);
    Integer potentialUser2ToUser1 = cardsRepository.
countMatchBetweenUsers(userId2, userId1);
    Integer count = Math.min(potentialUser1ToUser2,
potentialUser2ToUser1);
    if (count > 0) {
        ArrayList<CardEntity> result1 = new ArrayList<>(
                cardsRepository.
tradeCardsBetweenUsers(userId1, userId2, count));
        useAllCardAvailable(userId2);
        ArrayList<CardEntity> result2 = new ArrayList<>(
                cardsRepository.
tradeCardsBetweenUsers(userId2, userId1, count));
        useAllCardAvailable(userId1);
        ...
    }
}
```

This solution may seem more convenient as it simplifies the code and ensures consistency. However, in most scenarios, it can be an overkill. This kind of isolation is very costly for the database engine and can cause database locks and contention, degrading the performance of our application. It is a powerful mechanism but should be used wisely. We will explain in more detail in *How it works* section.

How it works...

When we use the `@Transactional` annotation in a method, Spring Data JPA creates a transaction when the method is invoked. If the method completes without errors, it commits the transaction and changes are confirmed. If an exception is thrown, Spring Data JPA rollbacks the transaction, setting the data to its previous state.

The `@Transactional` annotation can be applied at the class level. Then, Spring Data JPA applies the same behavior to all methods in that class.

An important concept to understand is transaction isolation. Why do we need to care about it? The key concept is concurrency. In concurrent systems, such as web applications, there are multiple operations simultaneously. Taking our card trading example, let's figure something out. During special events, there can be thousands or millions of users exchanging their cards. What happens if, for example, while trading between Sara and Paul, it turns out that Paul has already exchanged his cards with Joanna? If we don't control this scenario, it can happen that Sara gives her cards to Paul and Paul gives nothing to Sara. As we saw in the exercise, we can use some business logic to control this situation, or we can use higher isolation levels in the transaction. I recommend using higher isolation levels only when strictly required, as it hurts performance in high concurrent transactions.

In the example above we used `Serializable` isolation, which is the highest level of isolation. There are more in Spring Data JPA:

- `Isolation.DEFAULT`: The default isolation level is determined by the underlying database.
- `Isolation.READ_UNCOMMITTED`: Allows dirty reads, non-repeatable reads, and phantom reads.
- `Isolation.READ_COMMITTED`: Prevents dirty reads but allows non-repeatable reads and phantom reads.
- `Isolation.REPEATABLE_READ`: Prevents dirty reads and non-repeatable reads but allows phantom reads.
- `Isolation.SERIALIZABLE`: Provides the highest level of isolation, preventing dirty reads, non-repeatable reads, and phantom reads.

Keep in mind that the implementation of the isolation level relies on the underlying database engine, and there are some levels that may not be supported.

To define the Isolation levels, I used some terms that is worth explaining in detail:

- **Dirty Reads**: A dirty read occurs when one transaction reads data that has been modified by another transaction but not yet committed. In our example, it could be cards that were available for one user, that are no longer available once they have been exchanged.

- **Non-Repeatable Reads** (Uncommitted Data): Non-repeatable reads (or uncommitted data) occur when a transaction reads the same data multiple times during its execution, but the data changes between reads due to updates by other transactions.

- **Phantom Reads**: Phantom reads occur when a transaction reads a set of records that satisfy a certain condition, and then, in a subsequent read of the same records, additional records match the condition due to inserts by other transactions.

There is another concept that we haven't used in the example but is a core part of the Spring Data JPA. It is transaction propagation. What happens if a method annotated as @Transactional calls another @Transactional method? Are they executed in the same transaction or different ones? This behavior can be configured using the propagation attribute of @Transactional annotation. For instance:

```
@Transactional(propagation = Propagation.REQUIRES_NEW)
```

These are propagation possible values:

- REQUIRED (Default): If an existing transaction doesn't exist, a new transaction is started. If an existing transaction does exist, the method participates in that transaction.

- REQUIRES_NEW: A new transaction is always started, suspending any existing transactions. The method always runs in a new transaction, even if there was an ongoing transaction before.

- NESTED: Creates a "nested" transaction within the existing transaction, allowing for savepoints. If the nested transaction fails, it can roll back to the savepoint without affecting the outer transaction.

- NOT_SUPPORTED: The method runs without a transaction context. If an existing transaction exists, it's suspended while the method runs.

- MANDATORY: Requires an existing transaction to be present. If no transaction exists, an exception is thrown.

- NEVER: The method must not be run within a transaction context. If a transaction exists, an exception is thrown.

With propagation options, you can decide if you want to commit or rollback all changes, or you want to allow that some parts of the changes can be committed or rolled-back independently.

As you can see there many options related to transactions. I tend to use the default behavior and keep everything as simple as possible and use the available options when they are necessary.

There's more...

Here, we have used a declarative approach to implement transactions, but you can execute transactions with `EntityManager`, for instance when using Dynamic Queries.

```
em.getTransaction().begin();
// do your changes
em.getTransaction().commit();
```

Keep in mind that this way is more manual, so you need to properly manage the exceptions to rollback the transaction when needed. Usually, you will begin and commit your transaction inside a `try` block and you will rollback the transaction when an exception happens. That will look like this:

```
try {
    em.getTransaction().begin();
    ...
    em.getTransaction().commit();
} catch (Exception e) {
    em.getTransaction().rollback();
}
```

It is important to close the transactions as soon as possible, as depending on the level of isolation, they can lock data in the database and cause unwanted contention.

See also

It is possible to create distributed transactions involving more than one microservice, but doing so can be complex and comes with challenges. Distributed transactions that span multiple microservices require careful design and consideration of the distributed nature of microservices architectures. Traditional **two-phase commit** (**2PC**) is one way to achieve distributed transactions, but it's often avoided due to its complexity and potential for blocking and performance issues. Instead, many microservices architectures favor patterns like the Saga pattern or compensation-based transactions.

Here are some approaches for handling distributed transactions across multiple microservices:

- **Saga Pattern**: It is a way to maintain data consistency in a microservices architecture without relying on distributed transactions. In a saga, each microservice performs its part of the transaction and publishes events to inform other services about their actions. If an error occurs, compensating transactions are executed to revert previous actions. This pattern allows for eventual consistency and is often preferred in microservices.

- **Asynchronous Messaging**: Instead of tightly coupling microservices in a distributed transaction, you can use asynchronous messaging (e.g., with message queues) to communicate between services. Microservices can publish events when they complete their part of the work, and other services can consume these events and act accordingly.

- **Compensating Transactions**: In cases where something goes wrong, compensating transactions can be used to undo the changes made by a microservice. This is part of the Saga pattern and can help maintain data consistency.

- **API Gateway**: An API gateway can be used to orchestrate requests to multiple microservices as part of a single transaction. It can provide an API endpoint that aggregates multiple requests and enforces transactional semantics.

- **Distributed Transaction Coordinator** (**DTC**): While not commonly used in microservices architectures, you can implement a DTC that spans multiple microservices. However, this approach can introduce complexity and potential performance bottlenecks.

6

Data Persistence and NoSQL Database Integration with Spring Data

SQL and NoSQL databases offer a flexible and scalable approach to data storage and retrieval that can be better suited to certain use cases compared to traditional relational databases. NoSQL databases are designed for horizontal scale-out, flexibility, performance, high availability, and global distribution. However, with this, you lose the consistency, ACID compliance, and expressiveness of a full SQL implementation that a relational database can provide.

It's important to note that NoSQL databases are not a one-size-fits-all solution, and their suitability depends on the requirements of your application. In some cases, a combination of SQL and NoSQL databases might be the best approach to meet different data storage and retrieval needs within an organization.

In this chapter, we will use some of the most popular NoSQL databases. Each of them has a different approach to data access, but Spring Boot facilitates the developer experience in all of them.

First, we will learn how to use MongoDB, a document-oriented database that stores data in JSON-like objects. We will cover the basics of data access in MongoDB, as well as other advanced scenarios, such as indexing, transactions, and optimistic concurrency persistence.

Next, we will learn how to use Apache Cassandra. It is a wide-column store database, meaning that it stores data in tables with flexible schema and supports column-family data models. We will learn how to perform advanced queries, as well as how to manage optimistic concurrency persistency within it.

In this chapter, we will cover the following recipes:

- Connecting your application to MongoDB
- Using Testcontainers with MongoDB
- Data indexing and sharding in MongoDB
- Using transactions in MongoDB
- Managing concurrency with MongoDB
- Connecting your application to Apache Cassandra
- Using Testcontainers with Apache Cassandra
- Using Apache Cassandra templates
- Managing concurrency with Apache Cassandra

Technical requirements

For this chapter, you will need a MongoDB server and an Apache Cassandra server. In both cases, the easiest way to deploy them in your local environment is by using Docker. You can get Docker from its product page at `https://www.docker.com/products/docker-desktop/`. I will explain how to install MongoDB and Cassandra using Docker in the respective recipes.

If you wish to install MongoDB on your computer, you can follow the installation instructions on the product page: `https://www.mongodb.com/try/download/community`.

If you need to access MongoDB, you can use MongoDB Shell or MongoDB Compass, both of which can be found at `https://www.mongodb.com/try/download/tools`. I will be using MongoDB Shell in this chapter, so I recommend installing it.

For Cassandra, you can follow the instructions at `https://cassandra.apache.org/doc/latest/cassandra/getting_started/installing.html`.

All the recipes that will be demonstrated in this chapter can be found at: `https://github.com/PacktPublishing/Spring-Boot-3.0-Cookbook/tree/main/chapter6`.

Connecting your application to MongoDB

In this recipe, we will learn how to deploy a MongoDB server in Docker. Next, we will create a Spring Boot application and connect it to our MongoDB server using Spring Data MongoDB. Finally, we will initialize the database and perform some queries against the data that's been loaded.

We will use the scenario of football teams and players to demonstrate the different approaches to managing data in MongoDB compared to relational databases such as PostgreSQL.

Getting ready

For this recipe, we will use a MongoDB database. The easiest way to deploy it on your computer is by using Docker. You can download Docker from the product page at `https://www.docker.com/products/docker-desktop/`.

Once you have installed Docker, you can run a single instance of MongoDB or execute a cluster running in a replica set. Here, you will deploy a cluster running in a replica set. This is not necessary for this recipe but will be necessary for the following recipes as it is necessary to support transactions. I've prepared a script to facilitate the deployment of the cluster. This script deploys the cluster using `docker-compose`; once deployed, it initializes the replica set. You can find the script in this book's GitHub repository at `https://github.com/PacktPublishing/Spring-Boot-3.0-Cookbook/`, in the `chapter3/recipe3-2/start` folder.

You will need MongoDB Shell to connect to the MongoDB server. You can download it from `https://www.mongodb.com/try/download/shell`.

You will also need the *mongoimport* tool to import some data into the database. It is part of MongoDB's database tools. Follow the instructions on the product page to install it: `https://www.mongodb.com/docs/database-tools/installation/installation/`.

The data, once loaded, will look like this:

```
{
    "_id": "1884881",
    "name": "Argentina",
    "players": [
        {
            "_id": "199325",
            "jerseyNumber": 1,
            "name": "Vanina CORREA",
            "position": "Goalkeeper",
            "dateOfBirth": "1983-08-14",
            "height": 180,
            "weight": 71
        },
        {

            "_id": "357669",
            "jerseyNumber": 2,
            "name": "Adriana SACHS",
            "position": "Defender",
            "dateOfBirth": "1993-12-25",
            "height": 163,
            "weight": 61
        }
```

```
    ]
}
```

Each team has a list of players. Keep this structure in mind to better understand this recipe.

You can use *MongoDB Shell* to connect to the database. We will use it to create a database and initialize it with some data. You can find a script to load the data in this book's GitHub repository at `https://github.com/PacktPublishing/Spring-Boot-3.0-Cookbook/`. The script and the data are in the `chapter3/recipe3-1/start/data` folder.

How to do it...

Let's create a project using Spring Data MongoDB and create a repository to connect to our database. The database manages football teams, which include players. We will create some queries to get teams and players, and we will implement operations to make changes to our data. Follow these steps:

1. Create a project using the *Spring Initializr* tool. Open `https://start.spring.io` and use the same parameters that you did in the *Creating a RESTful API* recipe of *Chapter 1*, except change the following options:

    -   For **Artifact**, type `footballmdb`

    -   For **Dependencies**, select **Spring Web** and **Spring Data MongoDB**

2. Download the template that was generated with the *Spring Initializr* tool and unzip the content to your working directory.

3. First, we will configure Spring Data MongoDB to connect to our database. For that purpose, create an `application.yml` file in the `resources` folder. It should look like this:

    ```yaml
    spring:
        data:
            mongodb:
                uri:
    mongodb://127.0.0.1:27017/?directConnection=true
                database: football
    ```

4. Now, create a class named `Team` and annotate it with `@Document(collection = teams)`. It should look like this:

    ```java
    @Document(collection = "teams")
    public class Team {
        @Id
        private String id;
        private String name;
        private List<Player> players;
    }
    ```

Note that we also decorated the attribute ID with `@Id` and we are using `List<Player>` in our class. We will have a single collection of data in MongoDB called `teams`. Each team will contain players.

5. Next, create the `Player` class:

```
public class Player {
    private String id;
    private Integer jerseyNumber;
    private String name;
    private String position;
    private LocalDate dateOfBirth;
    private Integer height;
    private Integer weight;
}
```

The `Player` class does not require any special annotation as it's data that will be embedded in the `Team` document.

6. Now, create a repository to manage the teams persisted in MongoDB:

```
public interface TeamRepository extends MongoRepository<Team,
String>{
}
```

7. As it happens with `JpaRepository`, just by extending our `TeamRepository` interface from `MongoRepository`, we already have basic methods to manipulate `Team` documents in MongoDB. We will use this repository now. To do that, create a new service named `FootballService`:

```
@Service
public class FootballService {
    private TeamRepository teamRepository;

    public FootballService(TeamRepository teamRepository) {
        this.teamRepository = teamRepository;
    }
}
```

Now, we can create a new method in our service that retrieves a team using its `Id` value. This method in the service can use the `findById` method in `TeamRepository`, which is available by extending from `MongoRepository`:

```
public Team getTeam(String id) {
    return teamRepository.findById(id).get();
}
```

Spring Data MongoDB can automatically implement the queries of our `TeamRepository` interface just by following the standard naming convention. The details of the naming convention can be found at `https://docs.spring.io/spring-data/mongodb/reference/repositories/query-keywords-reference.html`. As an example, let's create a method to find a team by its name:

```
public Optional<Team> findByName(String name);
```

We can also create a method to find the teams with a name that contains a string:

```
public List<Team> findByNameContaining(String name);
```

8. Now, we will create a method to find a player. For this, we will need to look into the teams to find the player. It can be implemented by using the @Query annotation:

```
@Query(value = "{'players._id': ?0}", fields = "{'players.$':
1}")
public Team findPlayerById(String id);
```

9. As you can see, the query's `value` attribute is not SQL – it's in **MongoDB Query Language (MQL)**. Also, the query's `fields` attribute corresponds to the fields we want to retrieve from the document – in this case, just the `players` field of the document. This method will return a `Team` object with only one player.

Let's see how we can use this method. To do so, create a method in `FootballService` named `findPlayerById`:

```
public Player getPlayer(String id) {
    Team team = teamRepository.findPlayerById(id);
    if (team != null) {
        return team.getPlayers().isEmpty()
                                    ? null
                                    : team.getPlayers().get(0);
    } else {
        return null;
    }
}
```

We will use the `save` method of `MongoRepository` to *upsert* teams and `delete`/`deleteById` to make changes in the database:

*   **Upsert** means that if the data doesn't exist, this method will create the data, and will update the data if it already exists. Let's create a method named `saveTeam` in the `FootballService` class:

```
public Team saveTeam(Team team) {
    return teamRepository.save(team);
}
```

- Now, create a method to delete a team by its ID:

```
public void deleteTeam(String id) {
    teamRepository.deleteById(id);
}
```

In this recipe, we implemented a service that uses `MongoRepository` to perform the basic operations to interact with our MongoDB database. I've created a RESTful API to expose the methods that were implemented by the `FootballService` service that was created in this recipe. I've also created a script to make the requests to the RESTful API. You can find all this in this book's GitHub repository, in the `chapter6/reciper6-1/end` folder.

How it works...

When the application starts, Spring Data MongoDB scans the application looking for `MongoRepository` interfaces. Then, it generates the implementation for the methods defined in the repository and registers the interface implementation as a bean to make it available for the rest of the application. To infer the implementation of the interface, it uses the naming convention of the methods; see `https://docs.spring.io/spring-data/mongodb/docs/current/reference/html/#repository-query-keywords` for more details. Spring Data MongoDB also scans for methods with the `@Query` annotation to generate the implementation for those methods.

Regarding the `@Query` annotation, Spring Data MongoDB can do certain validations, but you should keep in mind that MongoDB is schema-flexible by design. This means that it doesn't assume that a field should or shouldn't exist. It will return a `null` value instead. Keep in mind that if the results are different from what you expected, there is probably a typo in your query.

In `findPlayerById`, we implemented a query to return an element of an array in a document. It is important to understand the data that's returned by MongoDB. When we want to find player `430530`, it returns the container document, a `Team` object with an `id` value of 1882891, with just the property players, and an array of just one element – that is, the player with ID `430530`. It looks like this:

```
[
  {
    "_id": "1882891",
    "players": [
      {
        "_id": "430530",
        "jerseyNumber": 2,
        "name": "Courtney NEVIN",
        "position": "Defender",
        "dateOfBirth": {
          "$date": "2002-02-11T23:00:00Z"
        },
```

```
            "height": 169,
            "weight": 64
        }
    ]
  }
]
```

> **Note**
>
> I've included players in the team collection for learning purposes. If you have a similar scenario and you expect to perform a lot of queries while searching for an element of an array in a collection, you might prefer to have a MongoDB collection for that array. In this case, I would store players in their own collection. It will perform and scale much better.

Here, `MongoRepository` provides three methods to save data:

- `save`: This method inserts a document if it doesn't exist and replaces it if it does exist. This behavior is also known as *upsert*.

- `saveAll`: This method behaves in the same way as `save`, but it allows you to persist a list of documents at the same time.

- `insert`: This method adds a new document to the collection. So, if the document already exists, it will fail. This method is optimized for inserting operations as it doesn't check the previous existence of the document.

The `save` and `saveAll` methods replace the document entirely if it already exists. If you only want to update certain properties of your entity, also known as partial document updates, you will need to use Mongo templates.

There's more...

I recommend looking to `MongoTemplate` for more advanced scenarios, such as when you need partial updates. Here's an example of it if you only want to update the team name:

```
public void updateTeamName(String id, String name) {
    Query query = new Query(Criteria.where("id").is(id));
    Update updateName = new Update().set("name", name);

mongoTemplate.updateFirst(query, updateName, Team.class);
}
```

As you can see, it allows you to define the `where` criteria to query the object and allows the *update* operation, defining which fields you want to update. Here, `MongoTemplate` is the core component used by Spring Data MongoDB to create the implementation of `MongoRepository` interfaces.

Using Testcontainers with MongoDB

When creating integration tests that depend on MongoDB, we have two options: using an in-memory database server embedded in our application or using Testcontainers. The in-memory database server can have slight differences from our production system. For that reason, I recommend using Testcontainers; it allows you to use a real MongoDB database hosted in Docker with all features enabled.

In this recipe, we'll learn how to set up a MongoDB Testcontainer and how to execute some initialization scripts so that we can insert test data into the database.

Getting ready

Executing Testcontainers requires a Docker-API compatible runtime. You can install Docker by following the instructions on the official web page: `https://www.docker.com/products/docker-desktop/`.

In this recipe, we'll add tests for the project we created in the *Connecting your application to MongoDB* recipe. I've created a working version of that recipe in case you haven't completed it yet. You can find it in this book's GitHub repository at `https://github.com/PacktPublishing/Spring-Boot-3.0-Cookbook`, in the `chapter6/recipe6-2/start` folder. In this folder, you will also find a file named `teams.json`. This will be used to initialize the data for the tests.

How to do it...

Let's enhance our project by creating automated tests with Testcontainers:

1. First, we'll need to include the Testcontainers dependencies. For that, open the `pom.xml` file and add the following dependencies:

```
<dependency>
    <groupId>org.testcontainers</groupId>
    <artifactId>junit-jupiter</artifactId>
    <scope>test</scope>
</dependency>
<dependency>
    <groupId>org.testcontainers</groupId>
    <artifactId>mongodb</artifactId>
    <scope>test</scope>
</dependency>
```

2. As we'll need to initialize the data in the database during the test execution, copy the `team.json` file described in the *Getting ready* section into the `tests/resources/mongo` folder.

3. Next, create a test class. Let's name it `FootballServiceTest`, and annotate the class with `@SpringBootTest` and `@Testcontainers`:

```
@SpringBootTest
@Testcontainers
class FootballServiceTest
```

4. We'll continue setting up the test class by creating the MongoDB container. As we'll see in the next step, we'll need to initialize the database with some data. For that, we'll copy the `teams.json` file described in *Step 2* to the container. We'll create the container and pass the file as follows:

```
static MongoDBContainer mongoDBContainer = new
MongoDBContainer("mongo").withCopyFileToContainer(
    MountableFile.forClasspathResource("mongo/teams.json"),
teams.json");
```

5. Now, we'll start the container and import the data from the `teams.json` file. To import the data, we'll use the *mongoimport* tool:

```
@BeforeAll
static void startContainer() throws IOException,
InterruptedException {
    mongoDBContainer.start();
    importFile("teams");
}
static void importFile(String fileName) throws IOException,
InterruptedException {
    Container.ExecResult res = mongoDBContainer.
execInContainer("mongoimport", "--db=football", "--collection="
+ fileName, "--jsonArray", fileName + ".json");
    if (res.getExitCode() > 0){
        throw new RuntimeException("MongoDB not properly
initialized");
    }
}
```

Note that this step should be performed before the tests start. That's why it's annotated with @ BeforeAll.

6. Now, we should configure the context so that it uses the MongoDB database hosted in Testcontainers. For that, we'll use the `@DynamicPropertySource` annotation:

```
@DynamicPropertySource
static void setMongoDbProperties(DynamicPropertyRegistry
registry) {
    registry.add("spring.data.mongodb.uri",
mongoDBContainer::getReplicaSetUrl);
}
```

7. Now that the MongoDB repository has been configured, we can continue with the normal test implementation. Let's inject `FootballService` into the test class and implement a simple test that will retrieve a `Team` object:

```
@Autowired
private FootballService footballService;
@Test
void getTeam() {
    Team team = footballService.getTeam("1884881");
    assertNotNull(team);
}
```

8. You can implement the tests for the rest of the functionality. I've created some basic tests for the `FootballService` class. You can find them in this book's GitHub repository at `https://github.com/PacktPublishing/Spring-Boot-3.0-Cookbook`, in the `chapter6/recipe6-2/end` folder.

How it works...

As we saw in the *PostgreSQL integration tests with Testcontainers* recipe of *Chapter 5*, by adding the `@Testcontainers` annotation, all the containers that are declared as static are available for all the tests in the class and stopped after the last test is executed. In this recipe, we used the specialized `MongoDBContainer` container; it provides the URL of the server, which we can use to configure the test context. This configuration is performed by using the `@DynamicPropertySource` annotation, as we saw in *Step 6*.

In this recipe, we learned how to copy files to the container and execute programs inside it. All the files in the `resources` folder are available at runtime. We copied the `teams.json` file to the container and then used the `mongoimport` tool to import the data into MongoDB. This tool is available in the MongoDB Docker image. One advantage of executing this tool in the container is that it's not necessary to specify the database server address.

Data indexing and sharding in MongoDB

In this recipe, we will manage football matches and their timeline – that is, the events that occur during the game. An event may involve one or two players and we must consider that the players' fans want to access all actions where their favorite players are involved. We will also consider that the number of matches and their events is growing every day, so we need to prepare our application to support all the load.

In this recipe, we'll introduce some key concepts to make your application performant and scalable. MongoDB, like relational databases, enables you to create indexes that optimize data access. If you plan to access certain data using the same parameters, it is worth creating indexes to optimize data read. Of course, you will require more storage and memory, and write operations will be impacted. So, you will need to plan and analyze your application needs.

As the size of your data increases, you will need to scale your MongoDB database. **Sharding** is a database architecture and partitioning technique that's used to horizontally partition data across multiple servers or nodes in a distributed system. With sharding, you can scale out your database by adding more servers and distributing the data across them using shards. A shard ensures that all data in the same shard will be on the same server.

In this recipe, we will use indexing and sharding in our football application while taking advantage of the features provided by Spring Data MongoDB. We will use other interesting features of Spring Data MongoDB, such as referring documents from other documents.

Getting ready

We will use the same tools that we did in the first recipe, *Connecting your application to MongoDB* – that is, Docker, MongoDB, and MongoDB tools such as *Mongo Shell* and *mongoimport*.

We will reuse the code from the *Connecting your application to MongoDB* recipe. If you haven't completed it yet, don't worry – I've prepared a working version in this book's GitHub repository at https://github.com/PacktPublishing/Spring-Boot-3.0-Cookbook/. It can be found in chapter6/recipe6-3/start. I've also created a script to load data into the database using the *mongoimport* tool. This can be found in the chapter6/recipe6-3/start/data folder.

The data is a bit different compared to what was provided in the *Connecting your application to MongoDB* recipe. I moved the players to their own MongoDB collection, and I added new collections to manage the matches and match events. If you want to keep the data from the previous recipe, I recommend that you create a new database for this recipe. You can do so by simply changing the --db parameter when calling the mongoimport tool. The calls will look like this:

```
mongoimport --uri="mongodb://127.0.0.1:27017/?directConnection=true"
 --db=football2 --collection=teams --jsonArray < teams.json
```

How to do it...

First, we will host the players' data in their own MongoDB collection. Players will be important entities for the new requirements, so they deserve their own collection. Then, we will create the document classes for matches and events. We will learn how to use Spring Data MongoDB annotations to configure MongoDB *indexes* and *shards*. Follow these steps:

1. Let's start by configuring players in their own MongoDB collection. Annotate the `Player` class with `@Document`:

    ```
    @Document(collection = "players")
    public class Player {
        @Id


        private String id;
    }
    ```

    Annotate the `id` field with the `@Id` annotation. We're doing this as it will be the document identifier.

    Now, remove the `players` field from `Team`.

2. Next, create the classes for matches and their events. For matches, we will create a class named `Match`:

    ```
    @Document(collection = "matches")
    public class Match {
        @Id
        private String id;
        private LocalDate matchDate;
        @Indexed
        @DBRef(lazy = false)
        private Team team1;
        @Indexed
        @DBRef(lazy = false)
        private Team team2;
        private Integer team1Goals;
        private Integer team2Goals;
    }
    ```

    Note that we started using two new annotations, `@Indexed` and `@DBRef`. They will be fully explained in the *How it works...* section of this recipe.

    For the match events, we will create a class named `MatchEvent`:

    ```
    @Sharded(shardKey = { "match" })
    @Document(collection = "match_events")
    public class MatchEvent {
    ```

```
@Id
private String id;
@Field(name = "event_time")
private LocalDateTime time;
private Integer type;
private String description;
@Indexed
@DBRef
private Player player1;
@Indexed
@DBRef
private Player player2;
private List<String> mediaFiles;
@DBRef
private Match match;
}
```

With that, we have introduced the `@Sharded` and `@Field` annotations.

3. To be able to use the new classes, we will create a repository for each class – that is, `PlayerRepository`, `MatchRepository`, and `MatchEventRepository`.

Let's look at `MatchEventRepository` in detail. It will implement the operations we need for our requirements:

*   Return all events in a match

*   Return all player events in a match:

```
public interface MatchEventRepository extends
MongoRepository<MatchEvent, String>{


    @Query(value = "{'match.$id': ?0}")
    List<MatchEvent> findByMatchId(String matchId);

    @Query(value = "{'$and': [{'match.$id': ?0}, {'$or':[
{'player1.$id':?1}, {'player2.$id':?1} ]}]}")
    List<MatchEvent> findByMatchIdAndPlayerId(String matchId,
String playerId);


}
```

4. At this point, we could run our application as Spring Data MongoDB components are in place. However, not all of the indexes have been created yet. If we want to create them as part of our application, we need to create a configuration class that extends `AbstractMongoClientConfiguration`, instructing Spring Mongo DB to create the indexes automatically:

```
@Configuration
public class MongoConfig extends
AbstractMongoClientConfiguration {
    @Override
    protected boolean autoIndexCreation() {
        return true;
    }
}
```

5. Now, we can create a service using these repositories to implement the new requirements of our application while connecting to MongoDB in an optimized way. I've created a service and a RESTful controller to demonstrate the use of these repositories. I've also added a few more tests using Testcontainers. You can find them in this book's GitHub repository at `https://github.com/PacktPublishing/Spring-Boot-3.0-Cookbook/`, in the `chapter6/recipe6-3/end` folder.

How it works...

First, I will explain the impact of the annotations that were used in this recipe.

The `@DBRef` annotation is a way to reference another document but keep in mind that this is a mechanism that's implemented by Spring Data MongoDB, not by the database engine itself. In MongoDB, the concept of reference integrity doesn't exist, and it should be managed at the application level. Here, `@DBRef` represents a document as an object with three fields:

- `$ref`: This contains the collection being referenced
- `$id`: This contains the ID of the document being referenced
- `$db`: This contains the database of the document being referenced

For example, here, you have a reference to team `1882891`:

```
{
    "$ref": "teams",
    "$id": "1882891",
    "$db": "football"
}
```

Spring Data MongoDB can use this annotation to automatically retrieve the referenced document. We can specify this behavior using the `lazy` attribute. By default, it is `true`, meaning that Spring Data MongoDB won't retrieve it automatically. If you set it to `false`, it will retrieve the referenced document automatically. We used this for the match document to automatically retrieve the information from the two teams playing the match.

The `@Indexed` annotation, as you may have figured out, creates an index in MongoDB. Then, the queries that use the indexed fields will perform faster for read operations.

The `@Sharded` annotation tells MongoDB how the collections should be distributed across shards. A server in a cluster can host one or more shards. We can also see a shard as a way to specify which documents will be hosted on the same server. In our case, we are interested in retrieving the events by match. This is the reason we configured `match` as the shard key. Selecting a good sharding key is crucial to make our application performant and scalable as it will impact the way the workload will be distributed across the servers.

When a query is performed in a sharded collection, MongoDB should identify if the request can be performed in a single shard or whether it will need to distribute the query across shards. It will gather the results from shards, aggregate them, and then return the results to the client. If you intentionally need to scale out a query horizontally, this is an excellent mechanism. It may happen that a request doesn't need to be distributed and could be executed in a single shard, but because of a wrong shard key selection, it is executed as a distributed query. The consequence is that it will consume more resources than expected because more servers will perform an unnecessary query, so the results have to be aggregated.

Sharding involves dividing a database into smaller parts, called shards, that can be hosted on a single server. A server can host multiple shards, and the shards are replicated across servers for availability. The number of servers can automatically increase or decrease, depending on the load. Sharding is useful for managing large datasets and large clusters, which are typically deployed in the cloud. For example, **MongoDB Atlas** can be hosted on cloud providers such as **Azure**, **Amazon Web Services** **(AWS)**, and **Google Cloud Platform** **(GCP)**, allowing the number of servers to be adjusted to meet real demand. However, in cases where the database is hosted in a single container on a computer, as in our example, sharding won't provide any significant benefits. In larger deployments, sharding is a crucial feature for achieving our goals.

We didn't explicitly create an index for `match` in `MatchEvent`, but it is implicitly created since it is the sharding key.

Finally, we used the `@Field` annotation. This is used to map a field in our document class to a different field in MongoDB. In our case, we mapped the `time` field in our class to the `event_time` field in MongoDB.

There's more...

Some decisions should be made when designing the data layer with MongoDB or other document-oriented databases. For example, should we mix different types of objects in the same collection, or should we keep each type of document in different collections?

Having different types of objects in the same collection can make sense if they share some common fields and you want to perform queries by those fields, or you want to aggregate data from different objects. For the rest of the scenarios, it is probably better to have each type of document in its own collection. It helps create indexes and facilitate the creation of shards.

In this recipe, we didn't mix different types of documents, but that is the reason why Spring Data MongoDB introduces a field named `_class` when it persists a document. For instance, this is the document that's persisted when creating a new team:

```
{
  "_id": "99999999",
  "name": "Mars",
  "_class": "com.packt.footballmdb.repository.Team"
}
```

Another decision to make is whether we should embed some data in a document or if that data should be in its own document. In the *Connecting your application to MongoDB* recipe, we embedded the players into their team, while in this recipe, we moved that information to its own collection. It may depend on the importance or independence of the embeddable document. In this recipe, the players required their own document as they can be directly referenced from other documents, such as match events.

There could be other reasons, such as the expected write concurrency over the embedded entity. For instance, we could embed the events in the matches. However, during a match, we could assume that there will be a high number of events happening. That operation would require a high number of write operations on the match document, which will require more consistency management.

Using transactions in MongoDB

We want to create a new service where users can purchase a virtual token that can be used to obtain virtual goods in this new game. The main goods are cards with player pictures and other information, a kind of virtual sticker.

There are two operations that we need to implement: the token purchase and the cards purchase. For the token purchase, there is a payment validation. Cards can only be purchased with tokens. Of course, the users will be able to purchase cards if they have enough tokens.

Since we need to ensure consistency regarding the token and cards balance, we will need to use transactions with our MongoDB repository.

In this recipe, we will learn more about MongoDB transactions and how they differ from relational database transactions.

Getting ready

We will use the same tools that we did in the *Connecting your application to MongoDB* recipe – that is, Docker and MongoDB.

We will reuse the code from the *Data indexing and sharding in MongoDB* recipe. If you haven't completed it yet, don't worry – I've prepared a working version in this book's GitHub repository at `https://github.com/PacktPublishing/Spring-Boot-3.0-Cookbook/`. It can be found in the `chapter6/recipe6-4/start` folder.

> **MongoDB transactions**
>
> MongoDB transactions are not supported in standalone servers. In the *Connecting your application to MongoDB* recipe, I provided a script to deploy a cluster using a replica set.
>
> In the *Deploying a MongoDB cluster in Testcontainers* recipe, we'll cover how to deploy multiple servers with containers to be able to test MongoDB transactions.

How to do it...

We will need to create a data model to support users and cards in our new service. Later, we will create a service that will use MongoDB transactions to perform operations involving users and cards consistently. We will configure our application so that it supports transactions as well. Follow these steps:

1. Let's start by creating the classes that will manage the objects to be stored in MongoDB:

    I. First, we will create a class named `User`:

    ```
    @Document(collection = "users")
    public class User {
        @Id
        private String id;
        private String username;
        private Integer tokens;
    }
    ```

    II. Next, we will create a class named `Card`:

    ```
    @Document(collection = "cards")
    public class Card {
        @Id
        private String id;
    ```

```
    @DBRef
    private Player player;
    @DBRef
    private User owner;
}
```

2. Next, we need to create the corresponding `MongoRepository` interfaces. Let's go:

I. Create an interface named `UserRepository`:

```
public interface UserRepository extends MongoRepository<User,
String>{
}
```

II. And another interface named `CardRepository`:

```
public interface CardRepository extends MongoRepository<Card,
String>{
}
```

3. Now, we need to create a service class that will manage the business logic of our application. To do that, create a class named `UserService`. Remember to annotate the class with `@Service`:

```
@Service
public class UserService {
}
```

4. This service will need the new repositories we created – that is, `UserRepository` and `CardRepository`, as well as `PlayerRepository`, which we created in the *Data indexing and sharding in MongoDB* recipe. We will need `MongoTemplate` as well. We will create a constructor with these repositories, after which the Spring Boot dependency manager will inject them:

```
@Service
public class UserService {

    private UserRepository userRepository;
    private PlayerRepository playersRepository;
    private CardRepository cardsRepository;
    private MongoTemplate mongoTemplate;

    public UserService(UserRepository userRepository,
                        PlayerRepository playersRepository,
                        CardRepository cardsRepository,
                        MongoTemplate mongoTemplate) {
        this.userRepository = userRepository;
```

```
        this.playersRepository = playersRepository;
        this.cardsRepository = cardsRepository;
        this.mongoTemplate = mongoTemplate;
    }
```

5. Next, we'll implement our business logic:

    I. Create a method for purchasing tokens named buyTokens:

```
public Integer buyTokens(String userId, Integer tokens) {
    Query query = new Query(Criteria.where("id").is(userId));
    Update update = new Update().inc("tokens", tokens);
    UpdateResult result = mongoTemplate.updateFirst(query,
update, User.class, "users");
    return (int) result.getModifiedCount();
}
```

    II. Create a method for purchasing cards named buyCards:

```
@Transactional
public Integer buyCards(String userId, Integer count) {
    Optional<User> userOpt = userRepository.findById(userId);
    if (userOpt.isPresent()) {
        User user = userOpt.get();
        List<Player> availablePlayers = getAvailablePlayers();
        Random random = new Random();
        if (user.getTokens() >= count) {
            user.setTokens(user.getTokens() - count);
        } else {
            throw new RuntimeException("Not enough tokens");
        }
        List<Card> cards = Stream.generate(() -> {
            Card card = new Card();
            card.setOwner(user);
            card.setPlayer(availablePlayers.get(
                    random.nextInt(0,
                            availablePlayers.size())));
            return card;
        }).limit(count).toList();
        List<Card> savedCards = cardsRepository.saveAll(cards);
        userRepository.save(user);
        return savedCards.size();
    }
    return 0;
}
```

6. To allow transactions in our application, we need to register a `MongoTransactionManager` bean. To do so, in our `MongoConfig` class, add the following method:

```
@Bean
MongoTransactionManager
transactionManager(MongoDatabaseFactory dbFactory) {
    return new MongoTransactionManager(dbFactory);
}
```

Now, our application can use transactions to execute operations atomically.

How it works...

By default, MongoDB native transactions are disabled in Spring Data MongoDB. That is the reason why we needed to register `MongoTransactionManager`. Once configured, when we annotate a method with `@Transactional`, it will create a transaction.

It is very important to note that transactions provide atomic operations, meaning that all of them are saved or none of them are, but they don't support isolation. The `buyCards` method will save all `cards` and the changes on `user` or it will save none of them.

An important difference compared to transactions in relational databases is that there is no locking or isolation. If we make changes to the same `User` document that is modified in `buyCards` in another request, it will raise a **write conflict exception**. MongoDB is designed for performance and scalability at the cost of losing features from ACID transactions. We'll learn how to manage concurrency in more detail in the *Managing concurrency with MongoDB* recipe.

As you've probably realized, the `buyTokens` method does not use transactions. The main reason is that it doesn't need to. All operations in a single document are considered isolated and atomic. Since the only field that's updated is `tokens`, we used the `inc` operation to modify the value. The advantage of this operator is that it's performed atomically in the server, even in high-concurrency environments. If we use transactions in operations involving a single document, it could raise write conflicts exceptions if two requests are updating the same document. This behavior could seem counter-intuitive if you compare it to the behavior of transactions in relational databases.

See also

In addition to `$inc`, there are other atomic operations in MongoDB worth knowing about for concurrent scenarios. They can be applied to fields and arrays. See `https://www.mongodb.com/docs/v7.0/reference/operator/update/` for more details.

Deploying a MongoDB cluster in Testcontainers

MongoDB transactions are only supported in multiple server clusters. However, `MongoDBContainer`, as explained in the *Using Testcontainers with MongoDB* recipe uses a single server deployment. Therefore, we cannot use it for the integration tests of the new feature to buy cards as it requires transactions.

In this recipe, we'll learn how to set up multiple Testcontainers and configure a MongoDB cluster. With that, we'll be able to implement integration tests for the buy cards feature.

Getting ready

This recipe will implement the integration tests for the *Using transactions in MongoDB* recipe. If you haven't completed it yet, don't worry – I've prepared a version from which you can start this recipe. You can find it in this book's GitHub repository at `https://github.com/PacktPublishing/Spring-Boot-3.0-Cookbook`, in `chapter6/recipe6-5/start`.

How to do it...

In this recipe, we'll set up the MongoDB cluster using Testcontainers and test the features involving transactions. Let's get started!

1. Since the new buy cards feature is the, we'll create a new test class named `UserServiceTest` and set up everything in this class. Since it uses Testcontainers, we'll annotate the class with `@Testcontainers`:

    ```
    @SpringBootTest
    @Testcontainers
    class UserServiceTest
    ```

2. Next, we'll create the MongoDB cluster. It will consist of three MongoDB containers deployed in the same network:

    I.   Declare a `Network` static field. This class is part of the Testcontainers library and allows us to define a Docker network:

    ```
    static Network mongoDbNetwork = Network.newNetwork();
    ```

    II. Now, create three static `GenericContainer` fields with the following properties:

    *   Each field will have the latest `mongo` Docker image.

    *   Each field will have the same network.

    *   The three containers will expose port `27017`.

    *   Each container will have a different network alias: `mongo1`, `mongo2`, and `mongo3`.

- The three containers will start with a mongod command that initializes the MongoDB cluster, with the only difference being the binding IP hostname. Each container will use its network alias.

III.   Here, we have the first field, mongoDBContainer1:

```
static GenericContainer<?> mongoDBContainer1 = new
GenericContainer<>("mongo:latest")
    .withNetwork(mongoDbNetwork)
    .withCommand("mongod", "--replSet", "rs0", "--port",
"27017", "--bind_ip", "localhost,mongo1")
    .withNetworkAliases("mongo1")
    .withExposedPorts(27017);
```

IV.   The other fields, mongoDBContainer2, and mongoDBContainer3, are declared as mongoDBContainer1, except we must change mongo1 to mongo2 and mongo3, respectively.

3.   Now that the three MongoDB containers have been declared, the next step is to start the containers and initiate the MongoDB replica set. We need to execute the following MongoDB command in one of the servers:

```
rs.initiate({
    _id: "rs0",
    members: [
        {_id: 0, host: "mongo1"},
        {_id: 1, host: "mongo2"},
        {_id: 2, host: "mongo3"}
    ]})
```

I've created a utility method named buildMongoEvalCommand to format the commands so that they're ready to be executed in MongoDB. We'll execute the MongoDB replica set initialization before any test execution. For that, we'll use the @BeforeAll annotation:

```
String initCluster = """
                rs.initiate({
                  _id: "rs0",
                  members: [
                      {_id: 0, host: "mongo1"},
                      {_id: 1, host: "mongo2"},
                      {_id: 2, host: "mongo3"}
                  ]
                })
                """;
mongoDBContainer1.start();
mongoDBContainer2.dependsOn(mongoDBContainer1).start();
```

```
mongoDBContainer3.dependsOn(mongoDBContainer2).start();
mongoDBContainer1.
execInContainer(buildMongoEvalCommand(initCluster));
```

4. The last configuration step is setting the mongodb address in the application using the @
    DynamicPropertySource annotation:

```
@DynamicPropertySource
static void setMongoDbProperties(DynamicPropertyRegistry
registry) {
    registry.add("spring.data.mongodb.uri", () -> {
        String mongoUri = "mongodb://" + mongoDBContainer1.
getHost() + ":" + mongoDBContainer1.getMappedPort(27017) +
"/?directConnect=true";
        return mongoUri;
    });
}
```

5. Now, we can create our testing methods. For instance, we'll create a test for the buyCards
    method of the UserService class:

```
@Test
void buyCards() {
    User user = new User();
    user.setUsername("Sample user");
    User createdUser = userService.createUser(user);
    Integer buyTokens = 10;
    userService.buyTokens(createdUser.getId(), buyTokens);
    Integer requestedCards = 1;
    Integer cardCount = userService.buyCards(user.getId(),
requestedCards);
    assertThat(cardCount, is(requestedCards));
    // do more assert
}
```

Some code snippets have been simplified or omitted for clarity. You can find more details in
this book's GitHub repository at https://github.com/PacktPublishing/Spring-
Boot-3.0-Cookbook.

How it works...

The `MongoDBContainer` container, which is available as a module in the Testcontainers project, only works as a single server deployment. For that reason, instead of using `MongoDBContainer`, we used `GenericContainer`. After that, we adapted to Testcontainers so that we could set up the script explained in the *Getting ready* section of the *Connecting your application to MongoDB* recipe. To do so, we did the following:

- Created a Docker network.

- Deployed at least three MongoDB servers in containers.

- Initialized the MongoDB replica set. A replica set is a group of Mongo processes that work together to maintain the same dataset. We can consider this a cluster.

As you may have noticed, we used the `directConnection` setting when connecting to the MongoDB cluster. This setting means that we connect directly to one specific node of the cluster. When connecting to a replica set, normally, the connection string specifies all cluster nodes and the client connects to the most appropriate one. The reason we use `directConnection` is because the nodes can discover each other using the network alias. After all, they are in the same network and can use DNS names. However, the application we developed runs on our development computer, which hosts the containers, but it's in a different network and cannot find the nodes by name. If we were in the same network, the MongoDB connection string would look like this:

```
mongodb://mongo1:27017,mongo2:27017,mongo3:27017/
football?replicaSet=rs0
```

In this case, the client will connect to the appropriate node. To execute transactions, it's necessary to connect to the primary server. The application we developed may fail when performing these transactions because it's not connected to the primary server.

> **Note**
>
> The `buildMongoEvalCommand` method has been adapted from the original `MongoDBContainer` container from the `Testcontainer` project. You can find the original code at `https://github.com/testcontainers/testcontainers-java/blob/main/modules/mongodb/src/main/java/org/testcontainers/containers/MongoDBContainer.java`.

Managing concurrency with MongoDB

In this recipe, we will implement a feature to exchange player cards between users. Some cards are more difficult to get, which results in them having a higher demand. So, while many users try to find them, only one may get it. This is a scenario of high concurrency.

A user can exchange or buy another user's card using a certain number of tokens. The process we will implement consists of the following steps:

1. First, we need to check that the buyer has the tokens they promised.

2. Then, we'll subtract the number of tokens from the buyer and add them to the seller.

3. Finally, we will change the card owner.

MongoDB supports optimistic concurrency control through a document's versioning system. Each document has a version number (often called a *revision* or *version* field) that is incremented whenever the document is modified. When multiple clients attempt to update the same document simultaneously, the version numbers are used to detect conflicts, and the changes are rejected if there is a conflict.

We will add version support to `cards` and `users` as we need to control that the users haven't spent the tokens on another thing and that the cards are not exchanged with another user.

Getting ready

We will use the same tools that we used in the *Connecting your application to MongoDB* recipe – that is, Docker and MongoDB.

We will reuse the code from the *Deploying a MongoDB cluster in Testcontainers* recipe. If you haven't completed it yet, don't worry – I've prepared a working version in this book's GitHub repository at `https://github.com/PacktPublishing/Spring-Boot-3.0-Cookbook/`. It can be found in the `chapter6/recipe6-4/start` folder.

How to do it...

Let's add version control support to our `Card` and `User` documents and implement a card exchange transaction with optimistic concurrency control:

1. First, we will modify the classes involved in our feature so that they support optimistic concurrency. We'll do this by adding a new field annotated with `@Version`:

    I.   Modify the `User` class by adding a new `Long` field named `version`:

    ```
    @Version
    private Long version;
    ```

    II.   And the same `version` field to the `Card` class.

2. Next, we will create a new service named `TradingService`:

    ```
    @Service
    public class TradingService {
        private CardRepository cardRepository;
    ```

```
        private UserRepository userRepository;

        public TradingService(CardRepository cardRepository,
                              UserRepository userRepository) {
            this.cardRepository = cardRepository;
            this.userRepository = userRepository;
        }
    }
```

Here, `CardRepository` and `UserRepository` are added as dependencies in the constructor as we will need them to implement the card exchange business logic.

3. Now, we will create two methods to implement the business logic. One will be annotated with `@Transactional` to control that all changes are atomic, and another one to control concurrency exceptions:

-   The business logic method should look as follows:

```
@Transactional
private Card exchangeCardInternal(String cardId, String
newOwnerId, Integer price) {
    Card card = cardRepository.findById(cardId).orElseThrow();
    User newOwner =
userRepository.findById(newOwnerId).orElseThrow();
    if (newOwner.getTokens() < price) {
        throw new RuntimeException("Not enough tokens");
    }
    newOwner.setTokens(newOwner.getTokens() - price);
    User oldOwner = card.getOwner();
    oldOwner.setTokens(oldOwner.getTokens() + price);
    card.setOwner(newOwner);
    card = cardRepository.save(card);
    userRepository.saveAll(List.of(newOwner, oldOwner));
    return card;
}
```

-   The method to control concurrency should look like this:

```
public boolean exchangeCard(String cardId, String newOwnerId,
                            Integer price) {
    try{
        exchangeCardInternal(cardId, newOwnerId, price);
        return true;
    } catch (OptimisticLockingFailureException e) {
        return false;
```

```
        }
    }
```

With this mechanism, we can control the concurrency operations that are performed on our documents. You can now implement a RESTful API that will use this business logic. I've prepared a working sample in this book's GitHub repository at `https://github.com/PacktPublishing/Spring-Boot-3.0-Cookbook/`. It can be found in `chapter6/recipe6-6/end`.

How it works...

By adding the `@Version` annotation, the save operations not only check that the `id` value is the same, but also the field annotated with `version`. The generated query looks like this:

```
Query query = new
Query(Criteria.where("id").is(id).and("version").is(version));
Update update = new Update().set("tokens", value).inc("version", 1);
mongoTemplate.updateFirst(query, update, User.class);
```

If this operation fails, it throws an `OptimisticLockingFailureException` exception.

Depending on the business requirements, we could retry the operations or just abandon them, as we did in our scenario. If a user already sold the card you want, you should look for another one.

Since we needed to modify three different documents, we used a transaction. We used the `@Transactional` annotation for declarative transaction management. If we want to roll back changes that have been performed in that transaction, we need to throw an exception. That is why we let Spring Data MongoDB throw `OptimisticLockingFailureException` in the `exchangeCardInternal` method and capture it in `exchangeCard`.

Connecting your application to Apache Cassandra

In this recipe, we want to create a system that allows users to post comments related to matches, players, or match events. We decided to use Apache Cassandra due to its high scalability and low-latency capabilities.

In this recipe, we'll learn how to connect our Spring Boot application to an Apache Cassandra server using Spring Data for Apache Cassandra repositories.

Getting ready

For this recipe, we will use an Apache Cassandra database. The easiest way to deploy Apache Cassandra on your computer is by using a container hosted in Docker. You can perform this task by executing the following `docker` command:

```
docker run -p 9042:9042 --name cassandra -d cassandra:latest
```

This command will download the latest Apache Cassandra Docker image, if you don't have one yet on your computer, and will start a Cassandra server listening on port 9042.

After starting the server, you will need to create a **Keyspace**. A Keyspace is a logical aggregation of tables. To create a Keyspace, you can execute the following Docker command. It will execute a cqlsh script inside the container:

```
docker exec -it cassandra cqlsh -e "CREATE KEYSPACE footballKeyspace
WITH replication = {'class': 'SimpleStrategy'};"
```

You may need to wait a few seconds for the Cassandra server to finish initializing before creating the Keyspace.

How to do it...

Let's create a project with Apache Cassandra support. We will use the already familiar Spring Data concept of Repository to connect to Apache Cassandra:

1. First, we will create a new Spring Boot project using the *Spring Initializr* tool. As usual, open https://start.spring.io. We will use the same parameters we used in *Chapter 1*, in the *Creating a RESTful API* recipe, except we'll use the following parameters:

    - For **Artifact**, type footballcdb

    - For **Dependencies**, select **Spring Web** and **Spring Data for Apache Cassandra**

2. Next, we will create a class named Comment. This represents the data for our new feature.

    We need to annotate the class with @Table and the fields with @PrimaryKeyColumn if they form part of the primary key. We can use @Column if we want to map a field to a different column name in Cassandra:

    ```
    @Table
    public class Comment {
        @PrimaryKeyColumn(name = "comment_id", ordinal = 0, type =
    PrimaryKeyType.PARTITIONED, ordering = Ordering.DESCENDING)
        private String commentId;
        private String userId;
        private String targetType;
        private String targetId;
        private String content;
        private LocalDateTime date;

        public Set<String> labels = new HashSet<>();
    }
    ```

The Comment table will include the comment content, the date, and the user posting the comment. It will also include information about the target of the comment – that is, a player, a match, or any other component we may have in our football application.

3. We need to create a new Repository for Comment that extends from CassandraRepository:

```
public interface CommentRepository extends
CassandraRepository<Comment, String>{
}
```

As is usual with Spring Data's Repository, it provides some methods to manipulate Comment entities, such as findById, findAll, save, and others.

4. As we will retrieve the comments when showing other entities, such as matches or players, we'll need to create a method in CommentRepository to get the comments by the type of target and the target itself:

```
@AllowFiltering
List<Comment> findByTargetTypeAndTargetId(String targetType,
String targetId);
```

Note that as with other repositories in Spring Data, it can implement the interface by inferring the query from the method name.

It is important to annotate the method with the @AllowFiltering annotation as we are not retrieving the data via the primary key.

5. We can now create a service using CommentRepository to implement our application requirements. We'll name the service CommentService and ensure it has the following content:

```
@Service
public class CommentService {
    private CommentRepository commentRepository;
    public CommentService(CommentRepository commentRepository){
        this.commentRepository = commentRepository;
    }
}
```

6. Now, we must create the functionality. We will create a method to create a comment and a couple of methods to retrieve all comments:

- We'll use a record to receive the comment data:

```
public record CommentPost(String userId, String targetType,
String targetId, String commentContent, Set<String> labels) {
}
```

- Let's define the `postComment` method so that we can create a new comment:

```
public Comment postComment(CommentPost commentPost) {
    Comment comment = new Comment();
    comment.setCommentId(UUID.randomUUID().toString());
    comment.setUserId(commentPost.userId());
    comment.setTargetType(commentPost.targetType());
    comment.setTargetId(commentPost.targetId());
    comment.setContent(commentPost.commentContent());
    comment.setDate(LocalDateTime.now());
    comment.setLabels(commentPost.labels());
    return commentRepository.save(comment);
}
```

- Now, we can create a method to retrieve all comments:

```
public List<Comment> getComments() {
    return commentRepository.findAll();
}
```

- We can retrieve all comments in general, but it makes more sense to retrieve the comments related to another entity. For instance, it's more common to get comments about a player:

```
public List<Comment> getComments(String targetType,
                                 String targetId) {
    return commentRepository.findByTargetTypeAndTargetId(
                                 targetType, targetId);
}
```

7. Finally, we need to configure the application so that it can connect to our Cassandra server. In the *Getting ready* section of this recipe, I provided instructions to deploy it on your computer by using Docker, including how to create a Keyspace. To configure the application, create an `application.yml` file in the `resources` folder. Add the following content:

```
spring:
    cassandra:
        keyspace-name: footballKeyspace
        schema-action: CREATE_IF_NOT_EXISTS
        contact-points: localhost

        local-datacenter: datacenter1
        port: 9042
```

8. We now have the components that are required to provide the comments functionality. We created `CassandraRepository` and we connected to a Cassandra server. I've created a RESTful API to consume this service. You can find it in this book's GitHub repository at `https://github.com/PacktPublishing/Spring-Boot-3.0-Cookbook/`. It can be found in `chapter6/recipe6-7/end`.

How it works...

As we saw with other Spring Data projects, when you create an interface extending from `CassandraRepository`, Spring Data for Apache Cassandra generates an implementation and registers the implementation as a *bean* to make it available for the rest of the components.

It can generate the implementation using the naming convention and using the **Cassandra Query Language (CQL)** with the `@Query` annotation. Both ways generate an implementation using Cassandra templates, something that will be detailed in the next recipe.

We haven't covered CQL yet, a syntactically similar language to SQL, but with important differences as Cassandra is a NoSQL technology. For instance, it doesn't support **JOIN** queries.

Note that in the `findByTargetTypeAndTargetId` method, we used `@AllowFiltering`. Cassandra is a NoSQL database that's designed for high availability and scalability but it achieves these features by limiting the types of queries it can handle efficiently. Cassandra is optimized for fast retrieval of data based on the primary key or clustering columns. When you query data in Cassandra, it's expected that you provide at least the primary key components to locate the data efficiently.

However, in some cases, you may need to perform queries that filter data on non-primary key columns. These types of queries are not efficient in Cassandra as they may require a full table scan and can be very slow on large datasets. You can use the `@AllowFiltering` annotation to explicitly indicate to Spring Data for Apache Cassandra that you're aware of the performance implications and that you want to perform such a query despite its potential inefficiency.

See also

It is recommended that you get familiar with CQL if you plan to work with Cassandra. You can find more information about it on the project page: `https://cassandra.apache.org/doc/stable/cassandra/cql/`.

Using Testcontainers with Cassandra

To ensure the reliability of our application, we need to run integration tests with our Cassandra project. Similar to MongoDB, we have two options for running tests with Cassandra – either by using an in-memory embedded Cassandra server or Testcontainers. However, I recommend using Testcontainers with a Cassandra server as this eliminates any potential compatibility issues since it uses a real Cassandra instance.

In this recipe, we will learn how to use the Testcontainers Cassandra module to create integration tests for our Comments service.

Getting ready

In this recipe, we will create an integration test for the Comments service that we created in the *Connecting your application to Apache Cassandra* recipe. If you haven't completed this recipe yet, you can use the project that I have prepared. You can find it in this book's GitHub repository at https://github.com/PacktPublishing/Spring-Boot-3.0-Cookbook/, in the chapter6/recipe6-8/start folder.

How to do it...

Are you ready to take your application to the next level? Let's start preparing it so that it can run Testcontainers and see how we can improve it!

1. We'll start by adding the Testcontainers dependencies to our pom.xml file – that is, the general Testcontainers dependency and the Cassandra Testcontainers module:

```
<dependency>
    <groupId>org.testcontainers</groupId>
    <artifactId>junit-jupiter</artifactId>
    <scope>test</scope>
</dependency>
<dependency>
    <groupId>org.testcontainers</groupId>
    <artifactId>cassandra</artifactId>
    <scope>test</scope>
</dependency>
```

2. Next, create a file named createKeyspace.cql in the test resources folder. This file should contain the Cassandra Keyspace creation command:

```
CREATE KEYSPACE footballKeyspace WITH replication = {'class':
'SimpleStrategy'};
```

3. Now, we can create a test class for our CommentService. You can name the test class CommentServiceTest. Before we start creating the test, we'll need to set up the Testcontainer. For that, do the following:

    I.   Annotate the test class with @Testcontainers:

```
@Testcontainers
@SpringBootTest
class CommentServiceTest
```

II.   Declare a static `CassandraContainer` field:

- Here, we'll specify the Cassandra Docker image. We'll use the default `cassandra` image.

- We must apply the Cassandra script to be executed during the container initialization process – that is, `createKeyspace.cql`, which we defined in *Step 2*.

- We must also expose the port where Cassandra listens for connections – that is, port 9042:

```
static CassandraContainer cassandraContainer =
(CassandraContainer) new CassandraContainer("cassandra")
            .withInitScript("createKeyspace.cql")
            .withExposedPorts(9042);
```

III. Before executing the tests, start the container. We'll use the `@BeforeAll` annotation for that purpose:

```
@BeforeAll
static void startContainer() throws IOException,
InterruptedException {
    cassandraContainer.start();
}
```

IV.   The last Testcontainers configuration involves setting the Cassandra connection setting in the application context. For that, we'll use `@DynamicPropertySource` and the properties that were provided by the `cassandraContainer` field we declared previously:

```
@DynamicPropertySource
static void setCassandraProperties(DynamicPropertyRegistry
registry) {
    registry.add("spring.cassandra.keyspace-name", () ->
"footballKeyspace");
    registry.add("spring.cassandra.contact-points", () ->
cassandraContainer.getContactPoint().getAddress());
    registry.add("spring.cassandra.port", () ->
cassandraContainer.getMappedPort(9042));
    registry.add("spring.cassandra.local-datacenter", () ->
cassandraContainer.getLocalDatacenter());
}
```

4.   Now, we can create our integration test. Let's name it `postCommentTest`:

```
@Autowired
CommentService commentService;
@Test
void postCommentTest() {
    CommentPost comment = new CommentPost("user1", "player",
"1", "The best!", Set.of("label1", "label2"));
    Comment result = commentService.postComment(comment);
```

```
            assertNotNull(result);
            assertNotNull(result.getCommentId());
    }
```

How it works...

The `org.testcontainers:cassandra` dependency contains the `CassandraContainer` class, which provides most of the functionality required to set up the Testcontainer for our integration test. It allows us to specify the Docker image we want to use.

Here, `withInitScript` executes CQL scripts in Cassandra by taking a file from the test's classpath. This simplifies execution as file copying and client tool availability are not a concern. We used this functionality to create the Keyspace, as we did in the *Getting ready* section of the *Connecting your application to Apache Cassandra* recipe.

We don't need to manually check if the Container service is ready to accept connections. Testcontainers automatically waits until the service is ready to initiate the tests.

Finally, we used the properties exposed by the `CassandraContainer` class to configure the connection. We used the `getContactPoint` method to get the server host address, the `getPort` method to get the port exposed by the container, and the `getLocalDatacenter` method to get the simulated datacenter name.

Using Apache Cassandra templates

We may want to access our data hosted in Cassandra in a more flexible way than the one provided by `CassandraRepository`. For instance, we may want to retrieve data from our comments system using a dynamic or complex query, execute operations in batch, or access a low-level feature. In those cases, it is more convenient to use a Cassandra template as it provides more low-level access to Cassandra's features.

In this recipe, we will implement functionality that will dynamically search comments using different parameters, such as a date range, tags, and so on. For that, we'll use **Cassandra templates**.

Getting ready

We will use the same tools that we did in the *Connecting your application to Apache Cassandra* recipe – that is, Docker and Apache Cassandra.

To complete this recipe, you'll need the project you created for the *Using Testcontainers with Cassandra* recipe. If you haven't completed that recipe yet, don't worry – you can use a full version of the project that I've prepared in this book's GitHub repository at `https://github.com/PacktPublishing/Spring-Boot-3.0-Cookbook/`. It can be found in the `chapter6/recipe6-9/start` folder.

How to do it...

In this recipe, we'll enhance the Comment service we created in the previous recipe with new search functionality so that users can use as many parameters as they want:

1. First, we need to inject `CassandraTemplate` into our `CommentService` class. To do that, modify the constructor so that it makes the Spring Dependency Container inject `CassandraTemplate`:

    ```
    @Service
    public class CommentService {
        private CommentRepository commentRepository;
        private CassandraTemplate cassandraTemplate;
        public CommentService(CommentRepository commentRepository,
                            CassandraTemplate cassandraTemplate) {
            this.commentRepository = commentRepository;
            this.cassandraTemplate = cassandraTemplate;
        }
    }
    ```

2. Now, add a new overloading for the `getComments` method:

    ```
    public List<Comment> getComments(String targetType,
                                    String targetId,
                                    Optional<String> userId,
                                    Optional<LocalDateTime> start,
                                    Optional<LocalDateTime> end,
                                    Optional<Set<String>> labels)
    ```

    This method has two types of parameters: mandatory and optional.

    We assume that users will always retrieve comments associated with a target entity – for instance, a player or a match. For that reason, the `targetType` and `targetId` parameters are mandatory.

    The rest of the parameters are optional; hence they are defined as `Optional<T>`.

3. In this new method, we will use the `QueryBuilder` component to create our query:

    ```
    Select select = QueryBuilder.selectFrom("comment").all()
                    .whereColumn("targetType")
                    .isEqualTo(QueryBuilder.literal(targetType))
                    .whereColumn("targetId")
                    .isEqualTo(QueryBuilder.literal(targetId));
    ```

    Here, we selected the comment table by using `selectFrom`, and we set the mandatory columns, `targetType`, and `targetId`, by using `whereColumn`.

The rest of the optional fields will use `whereColumn`, but only if they are provided:

```
if (userId.isPresent()) {
        select = select.whereColumn("userId")
                    .isEqualTo(QueryBuilder.literal(userId.
get()));
}
if (start.isPresent()) {
    select = select.whereColumn("date")
                    .isGreaterThan(QueryBuilder
                        .literal(start.get().toString()));

}

if (end.isPresent()) {
    select = select.whereColumn("date")
                    .isLessThan(QueryBuilder
                        .literal(end.get().toString()));

}
if (labels.isPresent()) {
    for (String label : labels.get()) {
        select = select.whereColumn("labels")
                    .contains(QueryBuilder.literal(label));

    }
}
```

4. Finally, we can use the query with `CassandraTemplate` by using the `select` method. Let's do it:

```
return cassandraTemplate.select(select.allowFiltering().build(),
                        Comment.class);
```

Here, we used `allowFiltering`. Since we are not using the primary key, we need to tell Cassandra that we assume that the query is potentially inefficient.

5. We implemented the new feature for our Comment service to perform dynamic queries using `CassandraTemplate`. Now, you can create a RESTful API interface to interact with the new feature. I've created a sample RESTful API that uses the new feature and prepared integration tests for the Comments service. You can find these in this book's GitHub repository at `https://github.com/PacktPublishing/Spring-Boot-3.0-Cookbook/`, in the `chapter6/recipe6-9/end` folder.

How it works...

Spring Data for Apache Cassandra registers a `CassandraTemplate` bean in the Spring Boot Dependency Container. It is used internally to implement the repositories described in the *Connecting your application to Apache Cassandra* recipe. By doing this, it can be injected into our components by Spring Boot.

You can compose a CQL string by concatenating the predicates, but this is prone to introducing typos in queries. That's why we used `QueryBuilder`. As I explained in the *Connecting your application to Apache Cassandra* recipe, we need to set `allowFiltering` when we make queries that don't use the table primary key.

There's more...

We can do the same query by building a string with a dynamic CQL statement. This would look like this:

```java
public List<Comment> getCommentsString(String targetType,
                                       String targetId,
                                       Optional<String> userId,
                                       Optional<LocalDateTime> start,
                                       Optional<LocalDateTime> end,
                                       Optional<Set<String>> labels) {
    String query = "SELECT * FROM comment WHERE targetType ='"
                 + targetType + "' AND targetId='" + targetId + "'";
    if (userId.isPresent()) {
        query += " AND userId='" + userId.get() + "'";
    }
    if (start.isPresent()) {
        query += " AND date > '" + start.get().toString() + "'";
    }
    if (end.isPresent()) {
        query += " AND date < '" + end.get().toString() + "'";
    }
    if (labels.isPresent()) {
        for (String label : labels.get()) {
            query += " AND labels CONTAINS '" + label + "'";
        }
    }
    query += " ALLOW FILTERING";
    return cassandraTemplate.select(query, Comment.class);
}
```

Managing concurrency with Apache Cassandra

We want to enhance our comments system by adding a new feature: upvoting comments. We will add a counter to our comments showing the positive votes that have been received.

This simple requirement can be complex in a high-concurrency scenario. If multiple users are upvoting a comment, it may happen that we aren't updating the latest version of the comment. To tackle this scenario, we will use an optimistic concurrency approach with Cassandra.

Getting ready

We will use the same tools that we did in the *Connecting your application to Apache Cassandra* recipe – that is, Docker and Apache Cassandra.

The starting point will be the project we created for the *Using Apache Cassandra Templates* recipe. If you haven't completed it yet, don't worry – you can use a full version of the project that I've prepared in this book's GitHub repository at `https://github.com/PacktPublishing/Spring-Boot-3.0-Cookbook/`. It can be found in the `chapter6/recipe6-10/start` folder.

How to do it...

In this recipe, we will implement the upvoting feature using optimistic concurrency. But before that, we'll need to prepare our comment entity. Let's get started:

1. The first thing we'll need to do is create a new field that will store the number of upvotes received by a comment. So, let's modify the `Comment` class by adding a new field named `upvotes`:

    ```
    private Integer upvotes;
    ```

2. We'll need to modify the table schema in the Cassandra server. For that, we'll need to connect to the server and execute a `cqlsh` command. The easiest way to do this is by connecting to the Docker container. The following command will open an interactive session in `cqlsh`

    ```
    docker exec -it cassandra cqlsh
    ```

    Ensure that `cqlsh` is in `footballKeyspace`. To do so, execute the following command in `cqlsh`:

    ```
    USE footballKeyspace;
    ```

    Now, alter the table comment to add the new `upvotes` column. For that, execute the following command in `cqlsh`:

    ```
    ALTER TABLE Comment ADD upvotes int;
    ```

    Now, you can exit `cqlsh` by executing the `quit;` command.

It isn't possible to assign default values in Cassandra. If you have existing comments in your database, Cassandra will return a `null` value for field upvotes. So, we'll need to manage this scenario accordingly.

3. Now, it's time to use the new field in a new operation. We'll implement that operation in our `CommentService` service by creating a new method named `upvoteComment`:

```
public Comment upvoteComment(String commentId) {
```

Next, we'll retrieve the first comment. We can use the existing `CommentRepository` or `CassandraTemplate`. We'll use `CommentRepository` as it is simpler:

```
Comment comment =
commentRepository.findByCommentId(commentId).get();
```

Now, we need to update the upvotes field, but we'll keep the current value:

```
Integer currentVotes = comment.getUpvotes();
if (currentVotes == null) {
    comment.setUpvotes(1);
} else {
    comment.setUpvotes(currentVotes + 1);
}
```

Next, we'll use the current value to create the condition. Only if we are updating the current value will we apply the change:

```
CriteriaDefinition ifCriteria = Criteria
                                .where(ColumnName.
from("upvotes"))
                                .is(currentVotes);
EntityWriteResult<Comment> result = cassandraTemplate
                                .update(comment,
                                    UpdateOptions.builder()
                                    .ifCondition(ifCriteria)
                                    .build());
```

Now, we need to check if the result was what we expected:

```
if (result.wasApplied()) {
    return result.getEntity();
}
```

If the result is not what we expected, we can retry the operation a few times while waiting a few milliseconds between executions, but this will depend on the requirements of the application.

Now, you can implement a RESTful API for this new functionality. I've prepared a sample RESTful API and integration tests in this book's GitHub repository at `https://github.com/PacktPublishing/Spring-Boot-3.0-Cookbook/`, in the `chapter6/recipe6-10/end` folder.

How it works...

The key to optimistic concurrency management in Cassandra is the conditional update command. In CQL, Cassandra provides an `IF` clause that we can use in `CassandraTemplate`. With this `IF` clause, you can conditionally update data, but only if certain conditions are met, which includes checking the current state of the data.

We could create a `version` field in the comments table to implement a mechanism, as we saw in the *Managing concurrency with MongoDB* recipe. However, Spring Data for Apache Cassandra does not provide any special capability to manage this automatically, so we would need to implement it ourselves. In addition, we don't expect any other change in the `comment` entity, so we can use upvotes to control if the row has been modified. The `upvotes` field is our `version` field.

Part 3: Application Optimization

In large-scale applications, it's necessary to understand where the bottlenecks are and how they can be improved. In this part, we'll follow a systematic approach to optimizing and measuring the improvements that we apply. We'll also use advanced techniques such as reactive programming and event-driven design.

This part has the following chapters:

- *Chapter 7, Finding Bottlenecks and Optimizing Your Application*
- *Chapter 8, Spring Reactive and Spring Cloud Stream*

7

Finding Bottlenecks and Optimizing Your Application

Finding what makes your application perform below your expectations can be difficult if you don't follow a systematic approach. When optimizing an application, it's important to focus your efforts on facts, not guesses. For that reason, in this chapter, we'll leverage the tools and learnings from *Chapter 3*, and we'll tackle some common challenges by analyzing the footprints of the changes applied.

In this chapter, you will learn how to use observability tools to find the bottlenecks of your application and apply some common techniques of application optimization, such as caching and runtime tuning. You will also learn how to improve your application's startup time and resource consumption by using native applications, which have been supported since Spring Boot 3's release.

We'll run some load tests to apply stress to our application, and we'll learn how to analyze the results.

In this chapter, we will go through the following recipes:

- Tuning the database connection pool
- Caching dependencies
- Using shared cache
- Using Testcontainers with Redis cache
- Creating a native image using Spring Boot
- Using GraalVM Tracing Agent to configure the native application
- Creating a native executable using Spring Boot
- Creating a native executable from a JAR

Technical requirements

I created an application that we'll optimize during this chapter. This application provides some RESTful APIs to manage football data. The application uses PostgreSQL as a data repository. You can find it on `https://github.com/PacktPublishing/Spring-Boot-3.0-Cookbook/`, in the `chapter7/football` folder. This application is already configured for observability, exposing a Prometheus endpoint with Actuator. To monitor the application, you can use Prometheus and Grafana.

> **Prometheus configuration**
>
> You will need to configure Prometheus, as explained in the *Integrating your application with Prometheus and Grafana* recipe in *Chapter 3*. I have already prepared the `prometheus.yml` file. You will need to get the IP address of your computer and set the value in the `prometheus.yml` file.

I created a Grafana dashboard to monitor the application's performance. To make it, I used the following dashboard as a starting point and adapted it for our purposes: `https://grafana.com/grafana/dashboards/12900-springboot-apm-dashboard/`.

In addition to PostgreSQL, Prometheus, and Grafana, we'll also use Redis for some recipes. As usual, the simplest way to run all these services on your computer is using Docker. You can get Docker from the product page: `https://www.docker.com/products/docker-desktop/`. I will explain how to deploy each tool in its corresponding recipe.

You may need a tool to execute SQL scripts in PostgreSQL. You can use the `psql` command-line tool or the more user-friendly *PgAdmin* tool. You can check the *Connect your application to PostgreSQL* recipe in *Chapter 5* for more details.

I prepared some JMeter tests to generate some load over the application. You can download JMeter from the project website at `https://jmeter.apache.org`.

For some of the recipes related to native applications, you will need the **GraalVM** JDK. You can follow the instructions to install it from the official website at `https://www.graalvm.org/downloads/`.

All the recipes that will be demonstrated in this chapter can be found at: `https://github.com/PacktPublishing/Spring-Boot-3.0-Cookbook/tree/main/chapter7`

Tuning the database connection pool

Database connections are an expensive resource that can take some time when they're created for the first time. For that reason, Spring Boot uses a technique known as connection pooling. When a connection pool is used, the application doesn't create a direct connection to the database; instead, it requests an available connection to the connection pool. When the application doesn't need a connection, it returns it to the pool. The connection pool usually creates some connections at the start of the application. When the connections are returned to the pool, they are not closed but reused by other parts of the application.

A common challenge when operating applications is deciding on the connection pool size. If the size is too small, under a certain load, some requests will take longer as they wait for a connection to become available in the pool. If the connection pool is too large, it will waste resources in the database server, as open connections are expensive.

In this recipe, we'll learn how to monitor the database connection pool in a Spring Boot application using standard metrics and monitoring tools. We'll use the techniques and tools learned in *Chapter 3*.

Getting ready

In this recipe, you will optimize an application that I have already prepared for this purpose. You can find the application in the book's GitHub repository at `https://github.com/PacktPublishing/Spring-Boot-3.0-Cookbook/`, in the `chapter7/football` folder. I recommend copying the folder's content to your working directory, as we'll apply different optimizations over the base project on each recipe.

The application uses PostgreSQL as a database engine and is configured for monitoring using Zipkin, Prometheus, and Grafana. You can run all these dependent services in Docker; for that purpose, I have prepared a `docker-compose-base.yml` file that you can find in the `chapter7/docker` folder. You can run this `docker-compose-base.yml` file by opening a terminal in the directory containing the file and executing the following command:

```
docker-compose -f docker-compose-base.yml up
```

The Prometheus service has a configuration file named `prometheus.yml` that contains the application scrapping configuration. It points to my computer's IP, but you will need to change it to your IP configuration. You should configure the Prometheus data source and the *SpringBoot APM Dashboard*. See the *Integrating your application with Prometheus and Grafana* recipe in *Chapter 3*, for more details.

I have prepared a JMeter test to generate workload on the application. You can find it in `chapter7/jmeter/Football.jmx`. This test simulates a common use case for the sample Football Trading application. The test performs the following steps:

1. One user buys some cards.

2. Another user buys some cards.

3. Both users try to use the cards in their albums.

4. Then, the first user gets all available cards from the second user and vice versa, the second user gets all available cards from the first user.

5. Both users examine the players on the cards from the other user.

6. They trade between them their available cards.

The test has 10 threads running simultaneously with no think time between requests.

How to do it...

We'll launch the application and ensure we see the application metrics in Grafana. Ready to find the application bottleneck and optimize it? Let's go for it!

1. First, we'll start the application, and we'll check that we see the application metrics in Grafana. I'll assume you have already started all dependent services as explained in the *Getting ready* section:

   - Open Grafana at `http://localhost:3000`, then open the SpringBoot APM Dashboard.

   - Check that you can see data in the **Basic Statics** and **HikariCP Statics** sections.

2. Start the JMeter application and open the `football.jmx` file, which you can find in the `chapter7/jmeter` folder.

3. Execute the JMeter test and wait until it finishes. The test execution can take some minutes to complete:

   - During the execution of the test, check the connection metrics in the **HikariCP Statistics** section in Grafana.

- You will see that there are pending connections:

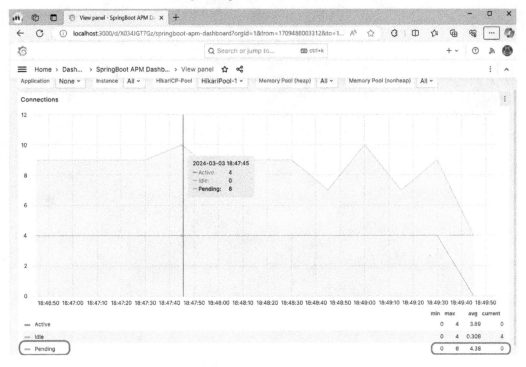

Figure 7.1: Hikari connection metrics

You can also see that the **Connection Acquire Time** value is over 4 ms all the time.

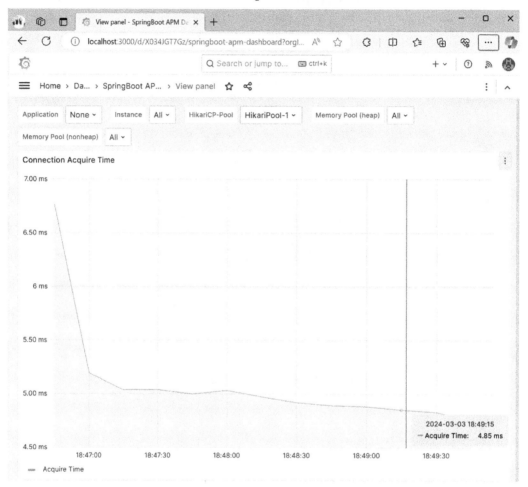

Figure 7.2: Connection Acquire Time

- You can see the results summary by opening the **Summary Report** item.

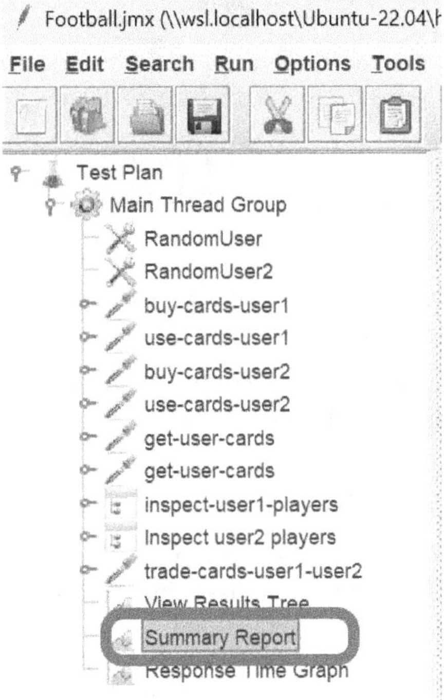

Figure 7.3: Summary Report

You can also see them while the test runs, but the baseline will be taken once completed.

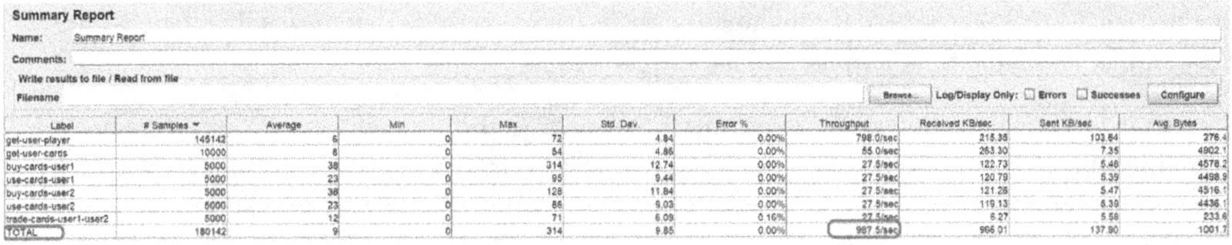

Label	# Samples	Average	Min	Max	Std. Dev.	Error %	Throughput	Received KB/sec	Sent KB/sec	Avg. Bytes
get-user-player	145142	6	0	72	4.84	0.00%	798.0/sec	215.35	103.64	276.4
get-user-cards	10000	8	0	54	4.86	0.00%	55.0/sec	263.30	7.35	4902.1
buy-cards-user1	5000	38	0	314	12.74	0.00%	27.5/sec	122.73	5.46	4578.2
use-cards-user1	5000	23	0	95	9.44	0.00%	27.5/sec	120.79	5.39	4498.9
buy-cards-user2	5000	38	0	128	11.84	0.00%	27.5/sec	121.28	5.47	4516.1
use-cards-user2	5000	23	0	86	9.03	0.00%	27.5/sec	119.13	5.39	4436.1
trade-cards-user1-user2	5000	12	0	71	6.09	0.16%	27.5/sec	6.27	5.56	233.6
TOTAL	180142	9	0	314	9.85	0.00%	987.5/sec	966.01	137.90	1001.7

Figure 7.4: Summary Report results – baseline results

In my environment, the total throughput is 987.5 **requests per second** (**RPS**), and the most used request is **get-user-player**, with a total of 145,142 requests and a throughput of 798 RPS. Note that the average time of the **get-user-player** operation is 6 milliseconds. Save the results of executing this test on your computer, as we'll compare them after the optimizations.

4. Now, we'll change the HikariCP settings by increasing the maximum number of database connections. For that, open the `application.yml` file in the `resources` folder and modify the `spring.datasource.hikari.maximum-pool-size` setting by increasing it to `10`.

5. Let's repeat the same performance test and see the difference. But before that, let's do a clean-up of the data to execute the test in the same conditions:

    I.   I prepared a script named `cleanup.sql` that you can run to clean up the database. You can find it in the `chapter7/dbscripts` folder.

    II. In JMeter, use the **Clear all** button to reset the results.

6. Once the test is done, compare the results with the baseline. The results on my computer are the following:

    - The total throughput is 1,315 RPS. That is approximately a 33% performance increase compared to the baseline 987.5 RPS.

    - The **get-user-player** request throughput is 1,085.3 RPS. That is approximately a 36% performance increase compared to the baseline 798 RPS.

    - The average response time of the **get-user-player** operation is 2 milliseconds. In the baseline, it was 6 milliseconds. That is three times faster.

    If you look at **HikariCP Statistics** in Grafana, you will see that there are no connections pending and the connection-acquire time has reduced. The connection acquire time metric on my computer is always below 10 microseconds.

How it works...

Spring Boot uses HikariCP as a JDBC data source connection pool. If you don't specify any pool size, the default is 10. I configured the initial example with a maximum of four connections for learning purposes. During the initial load test, we observed in Grafana that the number of pending connections remained consistently above zero throughout the entire testing period. That means that there is always a request that is waiting for an available database connection.

As we saw with the connection acquire time metric, on average, the time waited to acquire a connection is 4 milliseconds. That means that for every request, we need to add 4 milliseconds for each database operation involved. For fast operations, such as **get-user-player**, that is two times the time required when there's a connection available. Once we increased the size of the connection pool, this operation boosted its performance, and it was the most used operation in this scenario.

The rest of the operations also benefited from this new configuration, but as the request time with available connections is longer, the relative performance improvement is not that high.

In this recipe, we focused on the number of database connections. But the same approach can be applied to other types of application metrics, for instance, the number of Tomcat concurrent threads. You can use the observability data exposed by the application and tune your settings accordingly to adjust to your workload.

There's more...

In this recipe, we fixed the connection availability by adding the maximum number of connections used simultaneously at a given moment, that is, 10 connections. As mentioned, database connections are an expensive resource that should be used wisely. Let's consider a scenario with multiple instances of your service. Every additional connection for your application should be multiplied by the number of instances. Say you have 10 application instances; then, any additional connection should be multiplied by 10.

During the execution of the baseline test, we detected a maximum of six pending connections, so we added those six connections to the four initial connections. If the maximum number of pending connections happens only during a few spikes, we can adjust the number of maximum connections to 1 or 2 fewer connections than the maximum detected. For instance, in our scenario, we could adjust the number of maximum connections to 9, repeat the load test, and observe the impact.

Another potential adjustment is configuring the minimum and maximum number of connections. Then, if there is a spike and no available connections, HikariCP will create a connection to the database. Remember the time required to create the connection to the database and the time this connection will be idle. When the minimum and maximum connections are defined, HikariCP can close physical connections when idle. If the spike is too short, you may create a connection that will take longer than just waiting for an available connection, and then you will have an idle connection consuming resources in the database server.

Caching dependencies

The most common flow in the Football Trading application that we want to optimize is the following: sometimes, the users buy some cards, and after using them in their albums, they try to exchange the redundant cards they already have with other users. Before starting the exchange process, the users see which players are available from other users. There can be thousands and even millions of cards, but the total number of football players is around 700, and they are constantly retrieved from the Football Trading application.

Now, you want to optimize the application's performance. So, you are considering using a cache mechanism to avoid retrieving data from the database that is accessed frequently but changes very infrequently.

In this recipe, you will learn how to identify a database bottleneck and how to apply the caching mechanisms provided by Spring Boot. You will learn how to measure the improvement using the observability tools you learned about in *Chapter 3*.

Getting ready

In this recipe, you will continue optimizing the application that I prepared for this purpose. You can use the version resulting from the *Tuning the database connection pool* recipe. You can find the application in the book's GitHub repository at https://github.com/PacktPublishing/Spring-Boot-3.0-Cookbook/, in the chapter7/recipe7-2/start folder.

As we explained in the previous recipe, you can run all dependent services in Docker by running the docker-compose-base.yml Docker Compose file that you can find in the chapter7/docker folder. For that, open a terminal and execute the following command:

```
docker-compose -f docker-compose-base.yml up
```

We'll use the same JMeter test we used in the previous recipe. You can find it in chapter7/jmeter/football.jmx.

How to do it...

Let's start by executing the JMeter load test to determine the performance baseline. Then, we'll apply caching on different parts of the application, and we'll measure the improvements:

1. We can use the results of the JMeter execution from the *Tuning database connection pool* recipe test.

Summary Report

Name:	Summary Report							

Comments:

Write results to file / Read from file

Filename

Label	# Samples ▾	Average	Min	Max	Std. Dev.	Error %	Throughput	Rec
get-user-player	145683	2	0	41	1.25	0.00%	1085.3/sec	
get-user-cards	10000	5	0	57	2.12	0.00%	74.5/sec	
buy-cards-user1	5000	48	0	332	16.65	0.00%	37.1/sec	
use-cards-user1	5000	28	0	132	12.56	0.00%	37.2/sec	
buy-cards-user2	5000	47	0	159	15.68	0.00%	37.2/sec	
use-cards-user2	5000	28	0	109	12.70	0.00%	37.2/sec	
trade-cards-user1-user2	5000	13	0	157	8.57	1.98%	37.3/sec	
TOTAL	180683	7	0	332	12.72	0.05%	1340.3/sec	

Figure 7.5: JMeter summary report – baseline requests throughput detail

In my environment, the total throughput is 1,340.3 RPS, and the most used request is **get-user-player**, with a total of 145,683 requests and a throughput of 1,085.3 RPS. Save the results of executing this test on your computer, as we'll compare them after the optimizations.

2. Now that we have our application baseline, we'll enable caching:

    I. First, add the *Spring Cache Abstraction* starter to the pom.xml file:

```
<dependency>
    <groupId>org.springframework.boot</groupId>
    <artifactId>spring-boot-starter-cache</artifactId>
</dependency>
```

II.   Next, in the `FootballApplication` class, add the `@EnableCaching` annotation:

```
@EnableCaching
@SpringBootApplication
public class FootballApplication {
```

3. Next, we'll modify the `getPlayer` method of the `FootballService` class to cache the responses. This is the method called in the **get-user-players** step in JMeter. For that, you only need to annotate the method with `@Cacheable` as follows:

```
@Cacheable(value = "players")
public Player getPlayer(Integer id) {
    return playerRepository.findById(id).map(p -> playerMapper.
map(p)).orElse(null);
}
```

4. Let's execute the JMeter test again. But before that, let's do a cleanup of the data to execute the test in the same conditions:

I.   I prepared a script named `cleanup.sql` that you can run to clean up the database. You can find it in the `chapter7/dbscripts` folder.

II. In JMeter, use the **Clear all** button to reset the results.

5. Once the test is done, check the results and compare them with the baseline. The results on my computer are the following:

Summary Report

Label	# Samples ▾	Average	Min	Max	Std. Dev.	Error %	Throughput	Received KB/sec	Sent KB/sec	Avg. Bytes
get-user-player	148702	1	0	39	0.76	0.00%	1458.5/sec	393.84	189.43	276.4
get-user-cards	10000	8	0	36	1.59	0.00%	100.1/sec	480.99	13.38	4919.6
buy-cards-user1	5000	46	0	156	13.37	0.00%	50.1/sec	224.84	9.96	4599.5
use-cards-user1	5000	25	0	102	10.46	0.00%	50.1/sec	220.84	9.82	4516.8
buy-cards-user2	5000	46	0	196	13.94	0.00%	50.1/sec	222.34	9.96	4548.5
use-cards-user2	5000	25	0	104	10.71	0.00%	50.1/sec	218.49	9.82	4469.9
buy-cards-user2	5000	8	0	81	5.92	0.72%	50.1/sec	11.43	10.18	233.9
TOTAL	180702	6	0	196	12.13	0.02%	1806.6/sec	1770.12	262.21	1003.3

Figure 7.6: Summary report after applying caching on the FootballService

- The total throughput jumped to 1,806.7 RPS from 1,340.3 RPS, approximately a 34% performance increase.

- **get-user-player** requests are 1,458.5 RPS, and the baseline was 1,085.3 RPS, which means around a 34% performance increase as well.

- The rest of the requests also increased by around 34% of the overall throughput. For instance, **get-user-cards** rose to 100.1 RPS from 74.5 RPS, and the other requests went from 37.2 RPS to 50.1 RPS.

6. Let's use the caching in a different place in our application. Instead of applying the @Cacheable annotation in FootballService, apply the annotation in the PlayersController class in the getPlayer method:

```
@Cacheable(value = "players")
@GetMapping("/{id}")
public Player getPlayer(@PathVariable Integer id) {
    return footballService.getPlayer(id);
}
```

How it works...

By adding the *Spring Cache Abstraction* starter and using the @EnableCaching annotation, Spring Boot inspects the Beans for the presence of caching annotations on public methods, and a proxy is created to intercept the method call and handle the caching behavior accordingly; in our case, the methods annotated with @Cacheable. Spring Boot registers a CacheManager Bean to handle the cached items, as we didn't specify any specific CacheManager. Spring Boot uses the default implementation, a ConcurrentHashMap object, and it's handled in the process. This approach is valid for elements that do not change and where the dataset is small. Otherwise, you may want to use an external shared cache. In the next recipe, we'll tackle this scenario.

In this recipe, we optimized only get-user-player. It's the best candidate for all operations performed in this recipe. The reason is that the operations that modify data frequently are not candidates for caching, so buy-cards, use-cards, and trade-cards cannot be cached as they modify the data and are frequently used. The only operations that read just data are get-user-cards and get-user-player. get-user-cards is not a good candidate as the cards available owned by a user change every time they buy cards, exchange cards, or use them in an album. That means that the cache will be updated frequently. In addition, the number of users is high, around 100,000, so adding all those elements to the application memory can be counterproductive. On the other hand, get-user-player just retrieves the player's information. That information changes very infrequently, and there are just a few hundred players. For that reason, get-user-player is the best candidate for caching.

By adding the cache in the FootballService class, the throughput of that operation improved significantly, but it also benefited the rest of the operations. The reason is that even though it is a quick request on the database, it is the most frequent operation. The number of database connections available is defined by the hikaricp connection pool; we configured 10 connections. All operations should acquire a connection from hikaricp. As the most frequent operation is reduced, it's easier for the rest of the operations to acquire a connection faster.

There's more...

I recommend you check the metrics exposed by the application in Grafana while you run the tests. There are two main areas to observe in this scenario:

- **Basic statistics**: Here we can find the classic metrics for every application:

  - **CPU Usage**: This is often the limiting factor for demanding computing applications. During the tests on my computer, it was always under 70%.

  - **Heap Used**: This is the heap memory used by our application. It could limit the performance of our application.

  - **Non-heap Used**: This is all other memory used by our application. It usually accounts for less than 30% of the total memory used by the application, and its usage remains more stable than heap memory.

- **HikariCP Statistics**: As we saw in the previous recipe, HikariCP is the default database connection pool in Spring Boot. Creating a database connection to PostgreSQL or any other database engine is expensive. You can check the following metrics related to HikariCP:

  - **Active**: This is the number of connections out of the pool actively used to perform an operation in the database.

  - **Idle**: This is the number of available connections in the pool ready to be used when needed.

  - **Pending**: This is the number of operations waiting for an available connection to access the database. Ideally, this metric should be 0.

  - **Connection creation time**: This is the time spent creating the physical connection to the database.

  - **Connection usage time**: This is how long a connection is used before being returned to the pool.

  - **Connection acquire time**: This is the time required to get a connection. When there are idle connections, the time required will be very low. When there are pending connections, the time required will be higher.

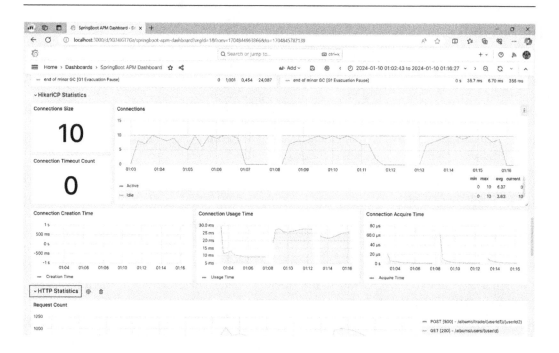

Figure 7.7: HikariCP metrics in Grafana

You may want to cache operations, as we did in this recipe, to reduce the number of connections to the database.

In the next recipe, we'll learn how to use Redis as an external cache and how to update it.

Using shared cache

The sample Football Trading application needs to cover a new scenario. Some football players can play in different positions, sometimes as defenders and sometimes as midfielders. Players do not change their position frequently, but it may happen. As we learned in the previous recipe, caching the players can improve the application's performance significantly. We assume it's possible and recommended to have more than one application instance running simultaneously. When a player is updated, all the application instances should return the latest version of the player.

In this recipe, we'll learn how to use an external cache shared among all application instances and how to update the cache when the underlying data is modified.

Getting ready

In this recipe, we'll reuse the application resulting from the previous recipe, as it has already been configured for caching. I have prepared a working version in the GitHub repository at `https://github.com/PacktPublishing/Spring-Boot-3.0-Cookbook/`. It is in the `chapter7/recipe7-3/start` folder.

The application uses PostgreSQL as a database engine, configured for observability with Zipkin, Prometheus, and Grafana.

As we'll add support for caching using Redis, we'll need a Redis server. The easiest way to run Redis on your computer is using Docker.

I prepared a Docker Compose file named `docker-compose-redis.yml`, with all dependent services, that is, PostgreSQL, Zipkin, Prometheus, Grafana, and Redis. You can find that file in the `chapter7/docker` folder. To run all dependent services, open a terminal in the `chapter7/docker` folder and run the following command:

```
docker-compose -f docker-compose-redis.yml up
```

I prepared a JMeter test to generate load for this recipe. You can find it in `chapter7/jmeter/Football-updates.jmx`. In addition to the flow implemented in the previous recipe, it updates the position of a football player from time to time.

How to do it...

We'll start by preparing the application to use Redis, and later, we'll ensure that the cache is updated when the players are modified:

1. First, we'll add the *Spring Data Redis* starter dependency. For that, just add the following dependency in the `pom.xml` file:

    ```
    <dependency>
        <groupId>org.springframework.boot</groupId>
        <artifactId>spring-boot-starter-data-redis</artifactId>
    </dependency>
    ```

2. You will need to add the following dependency to manage `LocalDate` fields:

    ```
    <dependency>
        <groupId>com.fasterxml.jackson.datatype</groupId>
        <artifactId>jackson-datatype-jsr310</artifactId>
        <version>2.16.1</version>
    </dependency>
    ```

3. Next, we need to configure Redis. For that, we will register a `RedisCacheConfiguration` Bean. Let's create a new configuration class; you can name it `RedisConfig`:

    ```
    @Configuration
    public class RedisConfig {
        @Bean
        public RedisCacheConfiguration cacheConfiguration() {
            ObjectMapper mapper = new ObjectMapper();
    ```

```
        mapper.registerModule(new JavaTimeModule());
        Jackson2JsonRedisSerializer<Player> serializer = new
Jackson2JsonRedisSerializer<>(mapper, Player.class);
        return RedisCacheConfiguration.defaultCacheConfig()
                .entryTtl(Duration.ofMinutes(10))
                .disableCachingNullValues()
                .serializeValuesWith(SerializationPair.
fromSerializer(serializer));
        }
    }
```

4. Finally, you must ensure that the cache is updated when the underlying data is updated. Let's modify the updatePlayerPosition method in the FootballService class by adding the @CacheEvict annotation:

```
@CacheEvict(value = "players", key = "#id")
public Player updatePlayerPosition(Integer id,
                                    String position)
```

5. Now, you can run the JMeter test to validate the application and measure the performance footprint. I prepared a test named Football-updates.jmx for that purpose. You can find it in the chapter7/jmeter folder. This test updates the player's position randomly but very infrequently, then retrieves the player to validate that it has the position updated.

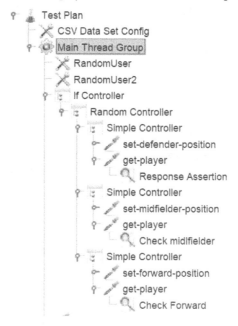

Figure 7.8: JMeter test, showing the details of the player update

On my computer, the total throughput is 1,497.5 RPS, and `get-user-players` is 1,210.6 RPS. The performance of Redis caching is slightly lower than in-process caching. However, externalizing the cache makes it possible to scale horizontally by adding additional instances.

How it works...

When adding an external caching implementation, the application needs to serialize the objects to be cached to send them over the network and save them in Redis. The default Redis configuration can manage basic types such as `String` or `int` with no additional configuration. However, in this sample application, we need to cache `Player` objects. To use the default configuration, the `Player` class should implement the `Serializable` interface.

To avoid modifying our domain classes, we configured a `Jackson2JsonRedisSerializer` serializer. This serializer represents the objects as JSON strings. The player has a catch with the `birthDate` field, as it is of the `LocalDate` type and cannot be managed with the default implementation. That is the reason we added the `com.fasterxml.jackson.datatype:jackson-datatype-jsr310` dependency and registered `JavaTimeModule` in `ObjectMapper` for `RedisCacheConfiguration`.

It's important to consider the implications of using an external cache repository:

- As we just learned, we must ensure the cached objects can be serialized.

- You need to consider the network latency as well. I executed all load tests locally on my computer, so there was no network latency. In real environments, it can also impact the performance of the application.

- The caching server may become the new bottleneck. Redis is very performant, but it may imply adding new resources to your solution, such as new servers.

I didn't notice significant performance differences in my load test results because everything ran on the same computer; however, you may expect slight differences in a production environment with services distributed across different servers.

You must configure the server address if you run Redis on a different server. By default, the *Spring Data Redis* starter assumes that Redis runs on `localhost` and listens to port `6379`.

In this recipe, we used the `@CacheEvict` annotation to update the cache. This annotation deletes the entry using a key. By default, this annotation uses all method parameters as the cache entry key. However, the `updatePlayerPosition` method has two parameters: the player `id` and the new `position`. As the key is just the player `id`, we specified that in the `position` field of the `@CacheEvict` annotation. Other options, such as clearing all entries, don't apply to our scenario.

Using Testcontainers with Redis cache

If you have executed the automated tests available in the sample project in the previous recipe, you may have noticed that the tests using methods that require Redis are failing. The reason is that Redis is not available during the execution of the tests.

In this recipe, we'll learn how to set up a Redis server hosted as a Docker container using Testcontainers.

Getting ready

In this recipe, we'll create the tests for the project created in the *Using shared cache* recipe. If you haven't completed it yet, use the version I prepared as a starting point for this recipe. You can find it in the book's GitHub repository at https://github.com/PacktPublishing/Spring-Boot-3.0-Cookbook, in the chapter7/recipe7-4/start folder.

As we use Testcontainers, you will need Docker installed on your computer.

How to do it...

We like reliable applications. Let's make our tests work!

1. We'll make all changes in the FootballServiceTest class. So, open it and add a new static field of type GenericContainer. We'll expose the default Redis port at 6379, and we'll use the latest redis image:

    ```
    static GenericContainer<?> redisContainer = new
    GenericContainer<>("redis:latest").withExposedPorts(6379);
    ```

2. Next, we'll modify the FootballServiceTest.Initializer class, by adding the properties to configure the connection to Redis:

    ```
    static class Initializer
                    implements
    ApplicationContextInitializer<ConfigurableApplicationContext> {
        public void initialize(ConfigurableApplicationContext
    configurableApplicationContext) {
            TestPropertyValues.of(
                "spring.datasource.url=" + postgreSQLContainer.
    getJdbcUrl(),
                "spring.datasource.username=" + postgreSQLContainer.
    getUsername(),
                "spring.datasource.password=" + postgreSQLContainer.
    getPassword(),
                "spring.data.redis.host=" + redisContainer.getHost(),
                "spring.data.redis.port=" + redisContainer.
    getMappedPort(6379))
                        .applyTo(configurableApplicationContext.
    ```

```
getEnvironment());
    }
}
```

3. Finally, start the container before executing the tests:

```
@BeforeAll
public static void startContainer() {
    postgreSQLContainer.start();
    redisContainer.start();
}
```

4. You can run the tests now. They should work fine!

How it works...

To integrate Redis into our tests, we only need an available Redis server. There is a specialized Redis module in Testcontainers. You can find it at `https://testcontainers.com/modules/redis/`. As the integration is pretty simple, we can use just `GenericContainer` instead of the specialized `RedisContainer`.

As we learned in previous recipes, by adding the `@Testcontainers` annotation in our test class, it automatically scans all container fields and integrates them into the test. `FootballServiceTest` was already annotated with `@Testcontainers` as it integrated PostgreSQL. We only needed to add a new container, in this case just `GenericContainer`, and perform the basic configuration to set it up. That is as follows:

- Configure the container with the minimum configuration: image and exposed port.

- Set the Redis configuration connection data in the application context. We did this in the `FootballServiceTest.Initializer` class. The Redis starter expects the configuration under `spring.data.redis`. We added the host and port, but only the port is required. By default, it expects the host on `localhost`.

- Start the container before the test execution. We did it in the method annotated with `@BeforeAll`.

Creating a native image using Spring Boot

Usually, when we design solutions using a microservice-oriented approach, we imagine that we can easily scale our applications by adding and removing new instances of our application, and we also imagine that this process happens immediately. However, starting new instances of our application can take longer than we initially expected. Spring Boot orchestrates Bean initialization, dependency injection, and event handling during application startup, and most of these steps happen dynamically. This is not a major issue for small applications, but for complex applications, this process can take up to minutes to complete.

Another important factor when designing applications is the efficient use of computing resources. We want the applications to consume as little memory as possible and process the workload efficiently.

For this kind of scenario, we can consider creating native applications, that is, applications that are built as final binaries for a specific processor family and operating system. A normal Java application generates intermediate code that is processed by the **Java Virtual Machine (JVM)** and converted into binary code during the application runtime. In a native application, this process happens during build time.

In this recipe, we'll learn how to create a new Spring Boot native application.

Getting ready

For this recipe, we'll need Docker. You can check the *Technical requirements* section of this chapter for more information.

How to do it...

Let's create a native application using Spring Boot!

1. Open the *Spring Boot Initializr* tool at `https://start.spring.io` and use the same options as you did in the *Create a RESTful API* recipe in *Chapter 1*, and use the same parameters, except the following options:

    *   For **Artifact**, type `footballnative`

    *   For **Dependencies**, select **Spring Web** and **GraalVM Native Support**

2. Next, create a sample RESTful controller; for instance, create a `TeamController` controller and a method that returns a list of teams:

    ```
    @RequestMapping("/teams")
    @RestController
    public class TeamController {
        @GetMapping
        public List<String> getTeams() {
            return List.of("Spain", "Zambia", "Brazil");
        }
    }
    ```

3. You can run the application on the JVM as usual, but what we'll do now is create a native Docker image. For that, open your terminal and execute the following Maven command:

    ```
    mvnw -Pnative spring-boot:build-image
    ```

Be patient, as this step can take up to a few minutes to complete depending on the resources of your computer.

4.   Once the build completes, you can run the Docker image with our native application by executing the following command in your terminal:

```
docker run --rm -p 8080:8080 footballnative:0.0.1-SNAPSHOT
```

5.   Now, you can perform a request to the RESTful application normally; for instance, you can use the following `curl` command:

```
curl http://localhost:8080/teams
```

How it works...

The GraalVM Native Support dependency adds a new `native` profile that can be used with the standard Spring Boot `build-image` goal to generate an image that targets **GraalVM**. GraalVM is a Java runtime that can compile your applications **Ahead of Time** (**AOT**) into native executables with low resource usage, fast startup, and improved security. To create the native image, the Maven plugin builds the native GraalVM executable in a Docker container using **Paketo Buildpacks**. Paketo Buildpacks are a set of community-driven tools that simplify the process of building and deploying applications as container images. That is the reason you don't need to download GraalVM tools on your computer.

The result is a Docker image that contains our application as a native executable. Just as a reference for performance improvements, the application takes around 1.5 seconds to start on my computer running on the JVM, while the native image takes 0.07 seconds to do the same. That is around 21 times faster. However, when running 10,000 requests, the total throughput of both versions is pretty similar, with the JVM version being a bit better performant. That could be because the native version runs on Docker, while the JVM runs directly on my computer. I prepared a JMeter test that you can use to compare the results on your computer. You can find a test named `teams-native.jmx` in the book's GitHub repository at `https://github.com/PacktPublishing/Spring-Boot-3.0-Cookbook`, in the `chapter7/jmeter` folder.

A native application is not a silver bullet that fits all scenarios. You need to consider that some features require dynamic processing during runtime, and they are difficult to handle with native applications. If your application doesn't have a quick boot-time requirement, native applications do not add many benefits and there can be many inconveniences. In terms of performance, a JVM application performs as well as a native application in the long term. That is, after warming up, it works as well as a native one; there can be some gains in terms of memory management, but in terms of performance, they are quite similar.

Using GraalVM Tracing Agent to configure the native application

The small native application we created in the previous recipe looks so promising that we've decided to build a bigger football application as a native app.

The application we created in the *Creating a native image using Spring Boot* recipe didn't require any special configuration. But native applications are built AOT. That means the compiler needs to analyze all code statically and detect the code reached during runtime. There are some Java technologies, such as the **Java Native Interface (JNI)**, **Reflection**, **dynamic proxy** objects, and class-path resources, that are very difficult to detect just by using static code analysis. The native compiler can use configuration files to include the required components in the final binary. As you may have figured out, the difficult part is configuring those files by detecting the components to be included in the final binary. For that purpose, GraalVM provides an agent that traces all usages to those types of technologies during the execution of an application on a regular JVM application.

In this recipe, we will build as a native image the sample application provided in this chapter. If you try to build the application as a native image as is, you will find some errors during runtime. In this recipe, we'll learn how to use GraalVM Tracing Agent to find all the required components and build a native image for an existing application. Then, you will be able to run your application in Docker.

Getting ready

In this recipe, we'll adapt the application you created in the *Using Testcontainers with Redis cache* recipe. If you haven't finished it yet, you can use a functional project I've provided as a starting point for this recipe. You can find it in the book's GitHub repository at `https://github.com/PacktPublishing/Spring-Boot-3.0-Cookbook` in the `chapter7/recipe7-6/start` folder.

You will need the GraalVM JDK installed on your computer. You can install it following the instructions from the official website at `https://www.graalvm.org/downloads/`.

The application depends on PostgreSQL, Redis, and other services. As we'll see in the *How to do it...* section, we'll run the application as a Docker container. To facilitate the execution in your development computer, I prepared a Docker Compose file named `docker-compose-all.yml` with the application and all dependent services.

How to do it...

Let's build a native executable image of our Spring Boot application. We'll see how fast it runs now! Remember that we initially created this application as a regular JVM application:

1. First, we'll add the *GraalVM Native Support* plugin to our pom.xml application. You should include the following configuration in the build/plugins element:

```
<plugin>
    <groupId>org.graalvm.buildtools</groupId>
    <artifactId>native-maven-plugin</artifactId>
</plugin>
```

2. Next, we'll need to add the Hibernate Enhance plugin as well. You should include the following configuration in the build/plugins element:

```
<plugin>
    <groupId>org.hibernate.orm.tooling</groupId>
    <artifactId>hibernate-enhance-maven-plugin</artifactId>
    <version>${hibernate.version}</version>
    <executions>
        <execution>
            <id>enhance</id>
            <goals>
                <goal>enhance</goal>
            </goals>
            <configuration>
                <enableLazyInitialization>true</
enableLazyInitialization>
                <enableDirtyTracking>true</enableDirtyTracking>
                <enableAssociationManagement>true</
enableAssociationManagement>
            </configuration>
        </execution>
    </executions>
</plugin>
```

3. In this step, we'll run the application using the GraalVM JVM with a special setting to trace the components that our application uses at runtime. The native compiler will use these traces to include those components in the final binary executable:

    *   This step requires that you use the GraalVM JVM. Depending on which installation method you used, switching Java versions may differ. I used the *SDKMAN!* tool, which just executes the following command in your terminal:

```
sdk use java 21-graalce
```

- To ensure you use the right JVM version, execute the following command in your terminal:

```
java -version
```

- Check that the response includes GraalVM. As a reference, this is the output when I execute this command on my computer:

```
bili@DESKTOP-GSIKJL4: ~          ×    +  ∨                                    ─   □   ×
bili@DESKTOP-GSIKJL4:~$ java -version
openjdk version "21" 2023-09-19
OpenJDK Runtime Environment GraalVM CE 21+35.1 (build 21+35-jvmci-23.1-b15)
OpenJDK 64-Bit Server VM GraalVM CE 21+35.1 (build 21+35-jvmci-23.1-b15, mixed mode, sharing)
bili@DESKTOP-GSIKJL4:~$ |
```

Figure 7.9: Sample java -version output for GraalVM JVM

- Build the application normally, that is, by executing the `package` goal in Maven. Execute this command on a terminal in the application root folder:

```
./mvnw package
```

- This command creates the JAR file for your application. By default, the filename will be `football-0.0.1-SNAPSHOT.jar`, and it will be created in the `target` directory.

- Now, run the GraalVM tracing tool. That is achieved by executing the application specifying an agent for the JVM, that is, specifying the `-agentlib:native-image-agent` parameter and passing the folder to save the configuration output. We'll set the folder where the native compiler expects the special configuration, that is, in `src/main/resources/META-INF/native-image`. This is how to execute your application specifying the GraalVM tracing tool:

```
java -Dspring.aot.enabled=true -agentlib:native-image-
agent=config-output-dir=src/main/resources/META-INF/native-image
-jar target/football-0.0.1-SNAPSHOT.jar
```

4. Now that our application is up and running, let's make sure that we cover all our bases. It's important that we execute every path of the application so that we can trace all the dynamic components and ensure everything is ready to build the native application. You'll see that the `src/main/resources/META-INF/native-image` folder contains several JSON files.

   When you complete the execution of all application paths, you can stop the application.

5. It's time to build the native application! You can do it by executing the following Maven command:

```
mvn -Pnative spring-boot:build-image
```

   The native build is way longer than the JVM one as it requires a deep static code analysis. It can take several minutes to complete. Be patient, it's worth the time!

6. Finally, you can run the application by executing a Docker container. As the application references other services that can no longer be located using `localhost`, you will need to specify some settings as environment variables. To make it easier for you, I've prepared a Docker Compose file. I named it `docker-compose-all.yml`, and you can find it in the book's GitHub repository.

On my computer, the native version takes just 1.29 seconds to be ready to accept requests, compared to 6.62 seconds for the JVM version.

How it works...

As explained in the *Creating a native image using Spring Boot* recipe, adding *GraalVM Native Support* to our application creates a new Spring Boot profile that we can use to build a Docker image with a native version of our application.

Some Hibernate operations generate Hibernate Proxy instances at runtime. If we don't include the Hibernate Enhance plugin, the native compiler doesn't have the required references at build time. For that reason, we need to include this plugin in our application.

In a simple application like the one created in the *Creating a native image using Spring Boot* recipe, we could skip *steps 3* and *4* and build the native application directly. However, we would realize that many operations do not work. That happens because the static build analysis doesn't detect some dynamic loading components, mostly related to Hibernate. To tackle this issue, GraalVM provides the Tracing Agent tool. This tool traces all usages of JNI, Java Reflection, dynamic proxy objects (`java.lang.reflect.Proxy`), or class-path resources and saves them in the specified folder. The files generated are the following:

- `jni-config.json`: This contains JNI-related information
- `reflect-config.json`: This contains reflection-related details
- `proxy-config.json`: This contains dynamic proxy object details
- `resource-config.json`: This contains class-path resource information
- `predefined-classes-config.json`: This contains metadata for predefined classes
- `serialization-config.json`: This contains serialization-related data

Then, the native compiler can use this configuration to include the referenced components in the final native executable. With this approach, we may find most of the components used at runtime, but some components may not be detected. In that case, we will need to include them manually.

As we run the application as a container, it's executed in the context of Docker. This means that to locate the dependent services, such as PostgreSQL, it's necessary to specify the internal Docker DNS name. In the previous recipes, all dependent services were accessible using `localhost`. For that reason, it's necessary to specify the address of all dependent components, for instance, by setting the environment variables, and the easiest way to set these environment variables is by creating a Docker Compose file.

There's more...

I executed the same JMeter tests we used in the *Using shared cache* recipe to compare the results of executing the same application running on a JVM and as a native application. In the following figure, you can see the results of running as a native application:

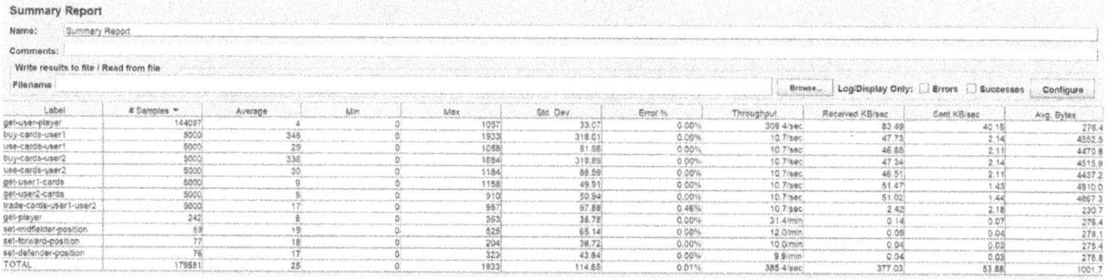

Summary Report

Name: Summary Report

Comments:

Write results to file / Read from file

Filename

Browse... Log/Display Only: Errors Successes Configure

Label	# Samples	Average	Min	Max	Std. Dev.	Error %	Throughput	Received KB/sec	Sent KB/sec	Avg. Bytes
get-user-player	144097	4	0	1057	33.07	0.00%	309.4/sec	83.49	40.18	276.4
buy-cards-user1	5000	346	0	1933	318.01	0.00%	10.7/sec	47.73	2.14	4552.5
use-cards-user1	5000	29	0	1058	81.68	0.00%	10.7/sec	46.88	2.11	4470.8
buy-cards-user2	5000	336	0	1694	318.89	0.00%	10.7/sec	47.34	2.14	4515.9
use-cards-user2	5000	30	0	1184	88.59	0.00%	10.7/sec	46.51	2.11	4437.2
get-user1-cards	5000	9	0	1158	49.91	0.00%	10.7/sec	51.47	1.43	4910.0
get-user2-cards	5000	9	0	910	50.94	0.00%	10.7/sec	51.02	1.44	4867.3
trade-cards-user1-user2	5000	17	0	957	67.88	0.46%	10.7/sec	2.42	2.18	230.7
get-player	242	8	0	363	38.78	0.00%	31.4/min	0.14	0.07	276.4
set-midfielder-position	89	19	0	526	65.14	0.00%	12.0/min	0.05	0.04	278.1
set-forward-position	77	18	0	204	38.72	0.00%	10.0/min	0.04	0.03	275.4
set-defender-position	76	17	0	323	43.64	0.00%	9.9/min	0.04	0.03	276.8
TOTAL	179581	25	0	1933	114.65	0.01%	385.4/sec	377.03	53.88	1001.7

Figure 7.10: JMeter throughput for native image running on Docker

The results may seem surprising, as the application running as a native application performs significantly worse than the JVM version.

There are two factors to keep in mind:

- The application now runs on Docker, while the application running on a JVM was executed directly on my computer
- Once the application running on a JVM did the **Just-in-Time** (**JIT**) compilation, there were no significant gains in performance compared to running

In the next recipe, we'll build the application natively instead of running on a container. Then, we'll be able to compare the applications running with similar conditions.

Creating a native executable using Spring Boot

In the previous recipes, we built the native application to run as a container. Even if that is a convenient solution for most modern cloud-native scenarios, we may need to build a native executable to be executed directly without a container engine.

In this recipe, we'll learn how to configure our computer to build native applications using the GraalVM JDK.

Getting ready

In this recipe, we'll reuse the result of the *Using GraalVM Tracing Agent to configure the native application* recipe. I prepared a version of the application that you can use as a starting point for this recipe. You can find it in the book's GitHub repository at `https://github.com/PacktPublishing/Spring-Boot-3.0-Cookbook/`, in the `chapter7/recipe7-6/start` folder.

You will need the GraalVM JDK version 21 installed on your computer. You can follow the instructions from the official website at `https://www.graalvm.org/downloads/`.

The application depends on some services, such as PostgreSQL and Redis. To facilitate the execution of these services on your computer, you can reuse the `docker-compose-redis.yml` file prepared in the *Using shared cache* recipe.

How to do it...

Now, we'll build our application as a native image that can be executed directly on our computer:

1. Ensure that you are using the GraalVM JVM for this process. For that, execute the following command:

    ```
    java -version
    ```

    Verify that the message contains GraalVM, as shown in *Figure 7.5*.

2. Next, we'll build the native executable. For that, open a terminal, change the directory to the root application folder, and execute the following command:

    ```
    mvn -Pnative native:compile
    ```

    The native build takes longer than a regular JVM build, even up to a few minutes.

3. Now, we have our binary executable in the `target` folder. Its name is the same as the project, this time without a version suffix. If you use Windows, it will be `football.exe`; in Unix-like systems, it will be just `football`. It's time to run the application. As I'm using Linux, I'll execute the following commands in my terminal:

    ```
    cd target
    ./football
    ```

    Be sure that the dependent services, such as PostgreSQL and Redis, are up and running. As explained in the *Getting ready* section, you can use the `docker-compose-redis.yml` Docker Compose file to run all dependent services.

How it works...

As we did in the *Using GraalVM Tracing Agent to configure the native application* recipe, we must prepare our application for the native build. In this recipe, we reused the application, and we already had the hints for the dynamic components that GraalVM needs to generate the native application. However, if you start from scratch, you will need to prepare the configuration as we did in the *Using GraalVM Tracing Agent to configure the native application* recipe.

The Spring Boot *GraalVM Native Support* starter includes the native profile and the `native:compile` goal. This starter was already included in the application we reused in this recipe. This time, the compilation process runs on your computer instead of being executed in a container.

There's more...

We can execute the load test using JMeter. This scenario is comparable to the one tested in the *Using Testcontainers with Redis cache* recipe, as both applications run directly on the computer and the dependent services run on Docker. These are the results of executing the same JMeter test on my computer:

Summary Report

Name: Summary Report

Comments:

Write results to file / Read from file

Filename

Label	# Samples ▼	Average	Min	Max	Std. Dev.	Error %	Throughput
get-user-player	145455	2	0	1019	12.81	0.00%	622.1/sec
buy-cards-user1	5000	165	0	1432	171.53	0.00%	21.4/sec
use-cards-user1	5000	16	0	698	31.76	0.00%	21.4/sec
buy-cards-user2	5000	159	0	1461	162.30	0.00%	21.4/sec
use-cards-user2	5000	16	0	918	34.19	0.00%	21.4/sec
get-user1-cards	5000	5	0	696	16.31	0.00%	21.4/sec
get-user2-cards	5000	5	0	807	26.09	0.00%	21.4/sec
trade-cards-user1-user2	5000	10	0	781	28.79	0.66%	21.4/sec
get-player	254	3	0	13	1.27	0.00%	1.1/sec
set-forward-position	91	8	0	15	2.59	0.00%	24.5/min
set-midfielder-position	85	8	0	16	2.78	0.00%	22.9/min
set-defender-position	78	8	0	16	2.74	0.00%	20.6/min
TOTAL	180963	12	0	1461	55.83	0.02%	773.5/sec

Figure 7.11: JMeter summary for a native application

The throughput for `get-user-player` is 622.1 RPS, compared to the 566.3 RPS achieved using a JVM version. That is approximately a 9.86% increase. For the total requests, it is 773.5 RPS compared to 699.2 RPS, which is approximately a 10.6% increase.

You must consider the benefits and trade-offs of using a native image. The main benefits are quick start-up time and better memory management and performance. The main trade-offs are the complexities of preparing the build image with all hints required to avoid runtime errors due to dynamic components. This configuration can be very painful and difficult to detect. You also need to consider the time required to build your application, which can be significantly longer than the JVM counterpart.

Creating a native executable from a JAR

As we realized during the completion of the previous recipes, building a native image takes way more time than building a regular JVM application. Another important consideration in certain environments is that GraalVM currently doesn't support cross-platform builds. That means if we need to build an application for Linux, as it's the most popular platform for server environments, but our development computer is a Windows or macOS computer, we cannot build the application directly. For these reasons, it could be a good choice to keep working with a regular JVM development process and create the native executable in a **Continuous Integration** (**CI**) platform. For instance, you can create a GitHub action for the native executable creation. In that way, we maintain the productivity for our development processes, we don't need to change our development platform, and we can target platforms for our application.

In this recipe, we'll generate a native executable for our football application using the *native-image* tool from the GraalVM JDK.

Getting ready

For this recipe, we'll use the outcome from the *Creating a native executable using Spring Boot* recipe. Creating a native executable using a *native-image* tool requires an AOT-processed JAR. If you plan to convert another application into a native executable, follow the instructions from the previous recipe to generate the AOT-processed JAR file. If you haven't completed the previous recipe yet, I prepared a working version that you can use as the starting point for this recipe. You can find it in the book's GitHub repository at `https://github.com/PacktPublishing/Spring-Boot-3.0-Cookbook/`, in the `chapter7/recipe7-8/start` folder.

You will need the *native-image* tool. This tool is part of the GraalVM JDK.

How to do it...

You can work normally with a JVM and keep the native build for CI. Let's see what you will need to do then!

1. The first step is ensuring that you generate a JAR with AOT processed. For that, open your terminal at the root of your project and package the JAR file using the `native` profile with Maven. To do so, execute the following command:

    ```
    ./mvnw -Pnative package
    ```

2. Next, we'll create a new directory for our native executable. Let's name it `native`. We'll create this directory inside the `target` directory:

    ```
    mkdir target/native
    ```

    Change your current directory to the new directory created:

    ```
    cd target/native
    ```

3. Now, we'll extract the classes from the JAR file created in *step 1*. We'll use the JAR tool, which is part of the JDK:

    ```
    jar -xvf ../football-0.0.1-SNAPSHOT.jar
    ```

4.   We can build the native application. For that, we'll use the `native-image` tool. We need to set the following arguments:

   - `-H:name=football`: This is the executable filename; in our case, it will be `football`.

   - `@META-INF/native-image/argfile`: The @ symbol indicates that the argument is read from a file. The specified file (`argfile`) likely contains additional configuration options or arguments for the native image generation process.

   - `-cp`: This argument sets the class path for the native image. We must pass the current directory, the `BOOT-INF/classes` directory, and all files contained in `BOOT-INF/lib`. This argument will look like this: `-cp .:BOOT-INF/classes:`find BOOT-INF/lib | tr '\n' ':'``.

   Then, to execute the `native-image` tool, you should execute the following command:

   ```
   native-image -H:Name=football @META-INF/native-image/argfile \
     -cp .:BOOT-INF/classes:`find BOOT-INF/lib | tr '\n' ':'`
   ```

5.   Now, you have our application built as a native executable. You can execute it just by executing the following command in your terminal:

   ```
   ./football
   ```

How it works...

As we reused the application from the previous recipes, we have already defined the hints. See the *Using GraalVM Tracing Agent to configure the native application* recipe for more details. To make them available for the native build, we must package our application using the `native` profile.

A JAR file contains the classes and resources of our application in a ZIP file. We could use a standard ZIP tool, but the JAR tool is more convenient for our purposes. We passed the `-xvf` arguments with the JAR file to be processed. The `x` argument instructs the tool to extract the content. `f` means that it will get the content from a file that is passed as an argument as well. Finally, `v` is just to generate a verbose output; we could get rid of this parameter.

For the `native-image` tool, we need to pass all files contained in the `BOOT-INF/lib` directory. Unfortunately, the cp argument doesn't admit wildcards. In Unix-like systems, you can use the `find` and `tr` tools. `find` lists the files in the directory, and `tr` removes \n and : characters. \n is the new line character.

8

Spring Reactive and Spring Cloud Stream

A different approach to applications may be necessary in high concurrency scenarios, with resource-intensive operations such as **Input/Output (I/O)**-bounded tasks that require low latency and responsiveness. In this chapter, we'll learn about two Spring Boot projects that address such a scenario.

Spring Reactive is Spring's response to reactive processing scenarios. Reactive processing is a paradigm that allows developers to build **non-blocking**, asynchronous applications that can handle **backpressure**. Non-blocking means that when an application waits for an external resource to respond, for instance, when calling an external web service or database, the application doesn't block the processing thread. Instead, it reuses the processing threads to handle new requests. Backpressure is a mechanism for handling situations where a downstream component cannot keep up with the rate of data production from an upstream component. For these mechanisms, Spring Reactive can be used in high-concurrency scenarios with resource-intensive operations.

Spring WebFlux is the reactive web framework equivalent to the Spring **model-view-controller (MVC)** we used in the previous chapters. To facilitate the transition between web frameworks, Spring WebFlux mirrors the names and annotations from the Spring MVC.

Spring Data **Reactive Relational Database Connectivity (R2DBC)** is the specification for integrating relational databases using reactive drivers. Compared to traditional blocking drivers, it also applies familiar abstractions.

Spring Cloud Stream is a framework for building highly scalable event-driven distributed applications connected with shared messaging systems. You can use reactive programming with Spring Cloud Stream, but the main goal of Spring Cloud Stream is to create loosely coupled distributed applications that can scale independently. Rather than trying to optimize the runtime execution as reactive does, Spring Cloud Stream provides the foundations for creating distributed applications that can work with some asynchronous degree. Spring Reactive and Spring Cloud Stream can be combined and complementary in high-concurrency scenarios.

In the first part of this chapter, we'll explore Spring Reactive by learning how to use Spring WebFlux and Spring Data R2DBC with PostgreSQL. In the second part, we'll learn how to use Spring Cloud Stream while using RabbitMQ as the messaging service. What you'll learn can be applied to other messaging services, such as Kafka, or other services provided by cloud providers, such as Amazon Kinesis, Azure Event Hub, or Google PubSub.

In this chapter, we're going to cover the following recipes:

- Creating a reactive RESTful API

- Using a reactive API client

- Testing reactive applications

- Connecting to PostgreSQL using Spring Data R2DBC

- Event-driven applications with Spring Cloud Stream and RabbitMQ

- Routing messages with Spring Cloud Stream and RabbitMQ

- Error handling with Spring Cloud Stream

Technical requirements

In this chapter, we'll need a PostgreSQL server and a RabbitMQ server. The easiest way to run them on your computer is by using Docker. You can get Docker from the official site at `https://docs.docker.com/engine/install/`. I will explain how to deploy each tool in its corresponding recipe.

All the recipes that will be demonstrated in this chapter can be found at: `https://github.com/PacktPublishing/Spring-Boot-3.0-Cookbook/tree/main/chapter8`.

Creating a reactive RESTful API

Spring Reactive is the Spring initiative that provides reactive programming features and capabilities that can be used in our Spring Boot applications. It is designed to support asynchronous and non-blocking programming. But what are asynchronous and non-blocking programming? To understand these concepts, it is better to start with the traditional model, the non-reactive programming model.

In a traditional model, when a Spring Boot application receives a request, a dedicated thread processes that request. If that request requires communicating with another service, such as a database, the processing thread is blocked until it receives a response from the other service. The number of available threads is limited, so if your application requires high concurrency but mostly waits for its dependent services to finish, this synchronous blocking model may have limitations.

In the reactive model, asynchronous and non-blocking programming reuses the threads across the concurrent requests and is not blocked by I/O operations, such as network calls or file operations.

Reactive programming is particularly well suited for building applications requiring high concurrency and scalability, such as web applications that handle many concurrent connections or real-time data processing systems.

In this recipe, we will use Spring WebFlux to build a RESTful API using reactive programming. Spring WebFlux is a module within Spring that enables reactive programming for building web applications.

Getting ready

This recipe doesn't have additional requirements. We will generate the project using the **Spring Initializr** tool and, once downloaded, you can make the changes with your favorite **Integrated Development Environment (IDE)** or editor.

How to do it...

In this recipe, we will create a RESTful API application. This time, we will create it using reactive programming, unlike the recipe in *Chapter 1*. Follow these steps:

1. Open `https://start.spring.io` and use the same parameters as in *Chapter 1*, in the *Creating a RESTful API* recipe, except changing the following options:

   - For **Artifact**, type `cards`

   - For **Dependencies**, select **Spring Reactive Web**

2. In the `cards` project, add a record named `Card`. Define the record as follows:

   ```
   public record Card(String cardId, String album,
                       String player, int ranking) {

   }
   ```

3. In the same folder, add a controller named `CardsController`:

   ```
   @RequestMapping("/cards")
   @RestController
   public class CardsController
   ```

   - Add a method named `getCards` that retrieves all cards:

   ```
   @GetMapping
   public Flux<Card> getCards() {
       return Flux.fromIterable(
               List.of(
                   new Card("1", "WWC23", "Ivana Andres", 7),
                   new Card("2", "WWC23", "Alexia Putellas", 1)));
   }
   ```

- And add another method to retrieve a card:

```
@GetMapping("/{cardId}")
public Mono<Card> getCard(@PathVariable String cardId) {
    return Mono.just(new Card(cardId, "WWC23", "Superplayer",
1));
}
```

In WebFlux, Flux<T> is used to return a stream of objects, while Mono<T> is used to return a single object. In non-reactive programming, they would be the equivalents to returning List<T> for Flux<T> and T for Mono<T>.

In this controller, Flux<Card> getCards() returns multiple objects of the Card type, and Mono<Card> getCard returns just one card.

4. Now, add an exception class named SampleException, implementing a new custom exception:

```
public class SampleException extends RuntimeException {
    public SampleException(String message) {
        super(message);
    }
}
```

5. Then, add two more methods to CardsController to demonstrate how to implement error handling in WebFlux:

```
@GetMapping("/exception")
public Mono<Card> getException() {
    throw new SampleException("This is a sample exception");
}


@ExceptionHandler(SampleException.class)
public ProblemDetail handleSampleException(SampleException e) {
    ProblemDetail problemDetail = ProblemDetail
                .forStatusAndDetail(HttpStatus.BAD_REQUEST,
                                    e.getMessage());
    problemDetail.setTitle("sample exception");
    return problemDetail;
}
```

The getException method always throws the exception, and handleSampleException handles exceptions of the SampleException type.

6. Now, open a terminal in the root folder of the cards project and execute the following command:

```
./mvnw spring-boot:run
```

We now have the RESTful API server running.

7. You can test the application by executing a request to `http://localhost:8080/cards`. You can use `curl` for this purpose:

```
curl http://localhost:8080/cards
```

You can also test how error handling works by requesting `http://localhost:8080/exception`. You will see that it will return an `HTTP 400` result.

How it works...

We used the same annotations as Spring Web to define the controllers in this recipe. However, the methods return the `Mono` and `Flux` types instead of traditional objects, indicating that the responses will be generated asynchronously. `Mono` and `Flux` are the core interfaces of the reactive programming model in WebFlux. `Mono` is used for asynchronous operations that produce at most one result, while `Flux` is used for asynchronous operations that return zero or more elements.

Reactive programming is centered around the concept of reactive streams. Reactive streams model asynchronous data flows with non-blocking backpressure. I mentioned some terms that may sound strange, so let me clarify them:

- **Non-blocking**: This refers to operations related to I/O, such as making an HTTP request, that avoid blocking threads. This enables the execution of a large number of concurrent requests without a dedicated thread per request.

- **Backpressure**: This is a mechanism to ensure that data is only produced as fast as it can be consumed, preventing resource exhaustion. For instance, this situation may happen when a downstream component cannot keep up with the data emitted by an upstream component. WebFlux manages the backpressure automatically.

There's more...

In addition to the annotation-based programming model used in this recipe, WebFlux also supports the **functional programming** model in defining the routes and handling requests. For instance, we can achieve the same result as the `cards` RESTful API with the following code:

1. First, create a class handling the logic:

```
public class CardsHandler {
    public Flux<Card> getCards() {
        return Flux.fromIterable(List.of(
                new Card("1", "WWC23", "Ivana Andres", 7),
                new Card("2", "WWC23", "Alexia Putellas",
1)));
    }
    public Mono<Card> getCard(String cardId) {
        return Mono.just(
```

```
                         new Card(cardId, "WWC23", "Superplayer", 1));
    }
}
```

2. And another one to configure the application:

```
@Configuration
public class CardsRouterConfig {
    @Bean
    CardsHandler cardsHandler() {
        return new CardsHandler();
    }
    @Bean
    RouterFunction<ServerResponse> getCards() {
        return route(GET("/cards"), req ->
            ok().body(cardsHandler().getCards(), Card.class));
    }
    @Bean
    RouterFunction<ServerResponse> getCard(){
        return route(GET("/cards/{cardId}"), req ->
            ok().body(
                cardsHandler().getCard(
                    req.pathVariable("cardId")), Card.class));
    }
}
```

Annotation-based programming is more like a traditional non-reactive programming model, while functional programming can be more expressive, especially for complex routing scenarios. The functional style is better suited for handling high concurrency and non-blocking scenarios because it naturally integrates with reactive programming.

Using annotation-based or functional is a matter of personal preference.

Using a reactive API client

We have a RESTful API, now it's time to use it in a non-blocking fashion. We'll create a reactive RESTful API that calls another RESTful API.

In this recipe, we'll create a reactive application that consumes an API. We'll learn how to use the reactive WebClient to perform requests to the target RESTful API.

Getting ready

In this recipe, we'll consume the application created in the *Creating a reactive RESTful API* recipe. If you haven't completed it yet, I prepared a working version that you can use as a starting point for this recipe. You can find it on the book's GitHub repository at `https://github.com/PacktPublishing/Spring-Boot-3.0-Cookbook`, in the `chapter8/recipe8-2/start` folder.

You can run the target project and keep it for the rest of the recipe.

How to do it...

We'll create an efficient consumer application for our RESTful API:

1. First, we'll create a new application using the Spring Boot Initializr tool. You can use the same options as in the *Creating a RESTful API* recipe in *Chapter 1*, except changing the following options:

    - For **Artifact**, type `consumer`

    - For **Dependencies**, select **Spring Reactive Web**

2. As we run the consumer application alongside the `cards` application, we'll need to change the port where the application listens for requests. We'll set `8090` as the server port. We will also create a configuration for the target football service URL. For that, open the `application.yml` file in the `resources` folder and set the following content:

    ```
    server:
        port: 8090
    footballservice:
        url: http://localhost:8080
    ```

3. Now, create a record named `Card` with the following content:

    ```
    public record Card(String cardId, String album,
                       String player, int ranking) {
    }
    ```

4. Then, we'll create a controller class named `ConsumerController` that will consume the target RESTful API. So, this controller will need a WebClient. For that, set the `ConsumerController` as follows:

    ```
    @RequestMapping("/consumer")
    @RestController
    public class ConsumerController {
        private final WebClient webClient;
        public ConsumerController(@Value("${footballservice.url}")
                                  String footballServiceUrl) {
    ```

```
        this.webClient = WebClient.create(footballServiceUrl);
    }
}
```

The controller now has a WebClient that allows us to perform requests in a non-blocking fashion in our client application.

5. Create a method to consume the operation from the other application that returns a stream of Card instances. For that, in ConsumerController, add the following method:

```
@GetMapping("/cards")
public Flux<Card> getCards() {
    return webClient.get()
            .uri("/cards").retrieve()
            .bodyToFlux(Card.class);
}
```

6. Create a method to consume the method returning a single object by adding the following method to the ConsumerController class:

```
@GetMapping("/cards/{cardId}")
public Mono<Card> getCard(@PathVariable String cardId) {
    return webClient.get()
            .uri("/cards/" + cardId).retrieve()
            .onStatus(code -> code.is4xxClientError(),
                    response -> Mono.empty())
            .bodyToMono(Card.class);
}
```

7. Then, create a method that manages different response codes from the remote server:

```
@GetMapping("/error")
public Mono<String> getFailedRequest() {
    return webClient.get()
            .uri("/invalidpath")
            .exchangeToMono(response -> {
                if (response.statusCode()
                        .equals(HttpStatus.NOT_FOUND))
                    return Mono.just("Server returned 404");
                else if (response.statusCode()
                        .equals(HttpStatus.INTERNAL_SERVER_
ERROR))
                    return Mono.just("Server returned 500: "
                            + response.bodyToMono(String.
class));
                else
```

```
                                      return response.bodyToMono(String.class);
                    });
        }
```

8. Let's run the consumer application now. When we perform requests to the client application, it will call the server RESTful API server application. Remember that we have the server RESTFul API server already running, as explained in the *Getting ready* section. Open a terminal in the root folder of the `consumer` project and execute the following:

    ```
    ./mvnw spring-boot:run
    ```

9. Now, test the `consumer` application. Remember that it listens on port 8090, and the server application listens on port 8080. In the terminal, execute the following command:

    ```
    curl http://localhost:8090/consumer/cards
    ```

    It will return a list of cards. The consumer application calls the server application to get the list of cards.

    Now, execute the following command:

    ```
    curl http://localhost:8090/consumer/cards/7
    ```

    It will return just a single `Card`. Again, the consumer application retrieved the `Card` with id number 7 from the RESTful API server application.

    ```
    curl http://localhost:8090/consumer/error
    ```

    It will return `Remote Server return 404`. The `consumer` application tried to call a method that does not exist in the server RESTful API server application. The consumer application handles the HTTP response codes from the server, in this case, `HttpStatus.NOT_FOUND` to return the final response message, which is `Remote Server return 404`.

How it works...

In this example, we consumed a RESTful API that is implemented using reactive technologies, but from the consumer's point of view, it doesn't matter. We can consume any RESTful API, regardless of the internal implementation.

What's important is that as we take advantage of a non-blocking client, the consumer application will benefit if it's also reactive. When we request against the consumer application, it will perform another request to the `cards` application. As we use a reactive client in the `consumer` application, it won't block a thread while the `cards` application responds, making that thread available to process other requests. This way, the application can manage higher concurrency than traditional blocking threads applications.

Testing reactive applications

As with non-reactive Spring Boot applications, we want to automate the testing of our reactive applications, and Spring Boot provides excellent support for testing such scenarios.

In this recipe, we'll learn how to create the tests using the components provided by default by Spring Boot when we add the **Spring Reactive Web** starter.

Getting ready

In this recipe, we'll create the tests for the projects used in the *Using a reactive API client* recipe. If you haven't completed that recipe yet, you can use the completed version that I prepared as a starting point for this recipe. You can find it in the book's GitHub repository at `https://github.com/PacktPublishing/Spring-Boot-3.0-Cookbook`, in the `chapter8/recipe8-3/start` folder.

How to do it...

We like robust and reliable applications. We'll do it with our Reactive applications:

1. As the applications from the *Using a reactive API client* recipe were created with the Spring Boot Initializr tool, just by adding Spring Reactive Web starter the testing dependencies are already included. You can check that the `pom.xml` file contains the following dependencies:

    ```
    <dependency>
        <groupId>org.springframework.boot</groupId>
        <artifactId>spring-boot-starter-test</artifactId>
        <scope>test</scope>
    </dependency>
    <dependency>
        <groupId>io.projectreactor</groupId>
        <artifactId>reactor-test</artifactId>
        <scope>test</scope>
    </dependency>
    ```

2. Now, we'll start the tests with the `cards` application. Create a new test class named `CardsControllerTest`. Remember, it should be created under the `test` folder; you can create it in the `src/test/java/com/packt/cards` folder:

    *   The test class should be annotated with `@WebFluxTest`:

        ```
        @WebFluxTest(CardsController.class)
        public class CardsControllerTests
        ```

- Then, we'll inject a `WebTestClient` field. For that, annotate the new field with `@Autowired`:

```
@Autowired
WebTestClient webTestClient;
```

- Now, we can use the `webTestClient` field to emulate the calls to the reactive RESTful API. For instance, let's create a test for the `/cards` path that returns a list of type `Card`. For that, create a new method annotated with `@Test`:

```
@Test
void testGetCards() {
    webTestClient.get()
            .uri("/cards").exchange()
            .expectStatus().isOk()
            .expectBodyList(Card.class);
}
```

- Let's test the `/cards/exception` path. For learning purposes, this path always returns a `404` code, a bad request result; and the body is of type `ProblemDetail`. The test method may look as follows:

```
@Test
void testGetException() {
    webTestClient.get()
            .uri("/cards/exception").exchange()
            .expectStatus().isBadRequest()
            .expectBody(ProblemDetail.class);
}
```

3. Next, we'll create the tests for the `consumer` application. As we want to test this application independently of the `cards` application, we'll need to mock the `cards` application server. As we learned in the *Mocking a RESTful API* recipe, in *Chapter 1*, we'll use the WireMock library. For that, open the `pom.xml` file of the project `consumer` and add the following dependency:

```
<dependency>
    <groupId>com.github.tomakehurst</groupId>
    <artifactId>wiremock-standalone</artifactId>
    <version>3.0.1</version>
    <scope>test</scope>
</dependency>
```

4. Now that we have all the dependencies, we'll create a new test class named `ConsumerControllerTest` and prepare it before writing the tests:

    I.   First, annotate the class with `@SpringBootTest` and set a few configuration options as follows:

```
@SpringBootTest(
        webEnvironment = SpringBootTest.WebEnvironment.RANDOM_
PORT,
        classes = {ConsumerApplication.class,
                ConsumerController.class,
                ConsumerControllerTests.Config.class})
public class ConsumerControllerTests
```

   II.   Note that we set a new class in the `classes` field that doesn't exist yet, `ConsumerControllerTests.Config`. It's used to configure the MockServer, as you'll see soon.

  III.   Next, we'll need to set up the WireMock server. For that, we'll create a configuration subclass class named `Config`; it will define a `WireMockServer` bean:

```
@TestConfiguration
static class Config {
    @Bean
    public WireMockServer webServer() {
        WireMockServer wireMockServer = new
WireMockServer(7979);
        wireMockServer.start();
        return wireMockServer;
    }
}
```

   IV.   Then, we need to configure the URI of the new remote server for the reactive WebClient. We need to set the `footballservice.url` application context variable. To perform this dynamic configuration, we'll use the `@DynamicPropertySource` annotation. For that, define a static method in the `ConsumerControllerTests` class:

```
@DynamicPropertySource
static void setProperties(DynamicPropertyRegistry registry) {
    registry.add("footballservice.url",
                () -> "http://localhost:7979");
}
```

V.   To complete the test preparation, we'll inject `WebTestClient` and `WireMockServer`, which we'll use in the tests. For that, define the fields with the `@Autowired` annotation:

```
@Autowired
private WebTestClient webTestClient;
@Autowired
private WireMockServer server;
```

5.   We can write the tests now. For instance, we'll create a test to get the cards:

I.   We can name it `getCards`:

```
@Test
public void getCards()
```

II.   First, we'll arrange what the mocked cards server will return. For that, we'll mock a small set of results for learning purposes:

```
server.stubFor(WireMock.get(WireMock.urlEqualTo("/cards"))
        .willReturn(
                WireMock.aResponse()
                .withStatus(200)
                .withHeader("Content-Type", "application/
json")
                .withBody("""
                [
                    {
                        "cardId": "1",
                        "album": "WWC23",
                        "player": "Ivana Andres",
                        "ranking": 7
                    },
                    {
                        "cardId": "2",
                        "album": "WWC23",
                        "player": "Alexia Putellas",
                        "ranking": 1
                    }
                ]""")));
```

III.   Then, we can perform the request using `webTestClient` and validate the results:

```
webTestClient.get().uri("/consumer/cards")
        .exchange().expectStatus().isOk()
        .expectBodyList(Card.class).hasSize(2)
        .contains(new Card("1", "WWC23", "Ivana Andres", 7),
                new Card("2", "WWC23", "Alexia Putellas", 1));
```

6.   You can write tests for the rest of the application's features. I created some sample tests, which you can find in the book's GitHub repository at `https://github.com/PacktPublishing/Spring-Boot-3.0-Cookbook`, in the `chapter8/recipe8-3/end` folder.

How it works...

With the `@WebFluxTest` annotation, we can define test classes focusing only on WebFlux-related components. This means that it will disable the configuration of all components save for those relevant to WebFlux. For instance, it will configure the classes annotated with `@Controller` or `@RestController`, but it won't configure classes annotated with `@Service`. With that, Spring Boot can inject `WebTestClient`, which we can use to perform requests to our application server.

In the consumer application, we need to mock the `cards` service. I won't go deeply into the details, as the mechanism is the same as explained in the *Mocking a RESTful API* recipe in *Chapter 1*. We used a configuration subclass annotated with `@TestConfiguration`. This annotation allows the configuration of beans that can be used alongside the tests. In our case, we just needed `WireMockServer`. Then, we dynamically configured the URI of the mocked server using the `@DynamicPropertySource` annotation.

> **Note**
>
> To reference the `Config` class, we used `ConsumerControllerTests.Config` instead of just `Config`. The reason for this is that it's a subclass of the `ConsumerControllerTests` class.
>
> We used the `webEnvironment` field, assigning `SpringBootTest.WebEnvironment.RANDOM_PORT`. This means that the test will host the application as a service on a random port. We used that to avoid port collisions with the remote server.

Connecting to PostgreSQL using Spring Data R2DBC

Using a Reactive database driver makes sense as we need to connect our Reactive application to PostgreSQL. This means the application is not blocked when it makes requests to the database. There is a Java specification to integrate SQL databases using reactive drivers named **R2DBC**. Spring Framework supports R2DBC with Spring Data R2DBC, which is part of the larger Spring Data family.

Spring Data R2DBC applies familiar Spring abstractions for R2DBC. You may use `R2dbcEntityTemplate`, running statements using the Criteria API and Reactive Repositories, among other features.

In this recipe, we'll learn how to connect to PostgreSQL using Reactive Repositories and some of the differences between Reactive and non-reactive Repositories. We'll also learn how to configure Flyway for database migrations.

Getting ready

For this recipe, we'll need a PostgreSQL database. You can use the instructions from the *Getting ready* section of the *Connecting your application to PostgreSQL* recipe, in *Chapter 5*. Once you have Docker installed, as explained in the aforementioned recipe, you can execute the following command to run a PostgreSQL server on Docker:

```
docker run -itd -e POSTGRES_USER=packt -e POSTGRES_PASSWORD=packt -p
5432:5432 --name postgresql postgres
```

I also prepared a starting project for this recipe that contains the classes that we'll use as data entities to map with the database tables and the database initialization scripts that we'll use in the *There's more* section for the Flyway migration. You can find the project in the book's GitHub repository at `https://github.com/PacktPublishing/Spring-Boot-3.0-Cookbook`, in the `chapter8/recipe8-4/start` folder.

How to do it...

We'll configure an application to connect to use PostgreSQL. Let's go Reactive:

1. First, we'll ensure we have all the required dependencies. For that, open the project's `pom.xml` file and add the following dependencies:

   - `org.springframework.boot:spring-boot-starter-webflux`

   - `org.springframework.boot:spring-boot-starter-test`

   - `io.projectreactor:reactor-test`

   - `org.springframework.boot:spring-boot-starter-data-r2dbc`

   - `org.postgresql:r2dbc-postgresql`

2. Next, we'll configure the connection to the database using the R2DBC driver. For that, open the `application.yml` file and add the following configuration:

   ```
   spring:
     application:
       name: football
   ```

```
    r2dbc:
        url: r2dbc:postgresql://localhost:5432/football
        username: packt
        password: packt
```

Note that the database URL doesn't start with jdbc: but with r2dbc:.

3. Then, we'll configure the entity classes we want to map to the database. These classes are in the repo folder. To prepare the classes, follow these steps for each of them:

    I.   Add the @Table annotation to the class. You can set the name as it's defined on the database.

    II.   Add the @Id annotation to the identifier field. I named this field Id in all entity classes.

    You can see the CardEntity class as an example here:

```
@Table(name = "cards")
public class CardEntity {
    @Id
    private Long id;
    private Optional<Long> albumId;
    private Long playerId;
    private Long ownerId;
}
```

4. We can create the repositories for our entities. For instance, for the CardEntity, we'll create the CardsRepository as follows:

```
public interface CardsRepository extends
                        ReactiveCrudRepository<CardEntity, Long>
{
}
```

    You can do the same for the rest of the entities.

5. We'll add a method to PlayersRepository to find a player by their name. For that, just add the following method definition to the PlayersRepository interface:

```
public Mono<PlayerEntity> findByName(String name);
```

6. Let's create a new service to manage players. You can name it PlayersService, and as it uses PlayersRepository, we'll add it as a parameter to the constructor, and we'll let Spring Boot do its magic to inject it:

```
@Service
public class PlayersService {
    private final PlayersRepository playersRepository;
    public PlayersService(PlayersRepository playersRepository) {
        this.playersRepository = playersRepository;
```

```
        }
    }
```

7. Now, we'll create a couple of methods using the repository. For instance, one method to get a player by the ID, and another one to get the player by the name:

```
public Mono<Player> getPlayer(Long id) {
    return playersRepository.findById(id)
            .map(PlayerMapper::map);
}
public Mono<Player> getPlayerByName(String name) {
    return playersRepository.findByName(name)
            .map(PlayerMapper::map);
}
```

Note that both methods use a class named `PlayerMapper`. I provided this class as part of the starting project to create the mappings between the entities and the objects returned by the application.

8. Let's make something more complex now. We'll retrieve a card and its related data, that is, `Album`, if it is already assigned, and `Player` in the card.

    I.   Let's create a new service class named `CardsService`. This service requires `CardsRepository`, `PlayersRepository`, and `AlbumsRepository`. We'll create a constructor with an argument of each type:

```
@Service
public class CardsService {
    private final CardsRepository cardsRepository;
    private final PlayersRepository playersRepository;
    private final AlbumsRepository albumsRepository;
    public CardsService(CardsRepository cardRepository,
                        PlayersRepository playersRepository,
                        AlbumsRepository albumsRepository) {
        this.cardsRepository = cardRepository;
        this.playersRepository = playersRepository;
        this.albumsRepository = albumsRepository;
    }
}
```

    II. Now, add a method to get an item of type `Card`:

```
public Mono<Card> getCard(Long cardId) {
    return cardsRepository.findById(cardId)
            .flatMap(this::retrieveRelations)
            .switchIfEmpty(Mono.empty());
}
```

III. As you can see in the getCard method, there is a reference to retrieveRelations. The retrieveRelations method retrieves Player from the database and Album in case it's defined. Of course, we'll do all this using a reactive approach:

```
protected Mono<Card> retrieveRelations(CardEntity cardEntity) {
    Mono<PlayerEntity> playerEntityMono =
            playersRepository.findById(cardEntity.getPlayerId());
    Mono<Optional<AlbumEntity>> albumEntityMono;
    if(cardEntity.getAlbumId() != null
        && cardEntity.getAlbumId().isPresent()){
        albumEntityMono = albumsRepository.findById(
                        cardEntity.getAlbumId().get())
                .map(Optional::of);
    } else {
        albumEntityMono = Mono.just(Optional.empty());
    }
    return Mono.zip(playerEntityMono, albumEntityMono)
            .map(tuple ->
                    CardMapper.map(cardEntity,
                            tuple.getT2(), tuple.getT1()));
}
```

9. You can implement the Reactive RESTful endpoints to expose this functionality, as explained in the *Creating a Reactive RESTful API* recipe in this chapter. I prepared some examples that you can find in the book's GitHub repository at https://github.com/PacktPublishing/Spring-Boot-3.0-Cookbook/ in the chapter8/recipe8-4/end folder. I recommend you check out the *There's more* section, as it includes Flyway to initialize the database and some tests using Testcontainers.

How it works...

Apparently, the Reactive entities and ReactiveCrudRepository are very similar to their non-reactive counterparts. ReactiveCrudRepository provides the same base methods with basic functionality such as findById and save, but there are important differences:

- The Reactive repositories don't manage relations between entities. For that reason, we haven't defined any @OneToMany or @ManyToOne fields. The relations between entities should be managed explicitly in our application, as we did in the getCard and retrieveRelations methods.

- The reactive repositories allow you to define methods following the same naming convention as the non-reactive repositories but returning Mono for single results and Flux for multiple results. These methods are transformed into queries in the database. You can find more details about the naming convention on the R2DBC web page at https://docs.spring.io/spring-data/relational/reference/r2dbc/query-methods.html.

- We haven't used it in this recipe, but it's possible to use the `@Query` annotation and provide an SQL query. It is a native query; JPQL is not supported.

The reactive programming model leverages the request's asynchronous and non-blocking process. Note that in the `getCard` method, the album and the player are retrieved asynchronously and simultaneously when the card is found. Parallelism is achieved using the `Mono.zip` method, which allows several non-blocking processes to be executed simultaneously.

There's more...

Flyway is not directly supported with R2DBC drivers, but it can be used with some adjustments. Let's see how to do it:

1. First, you must add the Flyway dependency in your `pom.xml` file:

```
<dependency>
    <groupId>org.flywaydb</groupId>
    <artifactId>flyway-core</artifactId>
</dependency>
```

2. As we want to validate the database migration automatically, we'll also include the `Testcontainers` support. For `Testcontainers`, there are no specific reactive adjustments to make.

```
<dependency>
    <groupId>org.testcontainers</groupId>
    <artifactId>junit-jupiter</artifactId>
    <scope>test</scope>
</dependency>
<dependency>
    <groupId>org.testcontainers</groupId>
    <artifactId>postgresql</artifactId>
    <scope>test</scope>
</dependency>
```

3. Now, we'll need to configure the connection for Flyway explicitly. The reason for this is that Flyway only supports the JDBC driver; for that reason, we'll need to specify the `jdbc:` version of the database URL. This configuration can be applied in the `application.yml` file. It should look like this:

```
spring:
    r2dbc:
        url: r2dbc:postgresql://localhost:5432/football
        username: packt
        password: packt
```

```yaml
  flyway:
    url: jdbc:postgresql://localhost:5432/football
    user: packt
    password: packt
```

We can set this configuration in a test supporting Testcontainers. Let's see what the class to test the PlayersService may look like:

```java
@Testcontainers
@SpringBootTest
@ContextConfiguration(initializers = PlayersServiceTest.
Initializer.class)
class PlayersServiceTest {
    @Autowired
    private PlayersService playersService;

    static PostgreSQLContainer<?> postgreSQLContainer = new
PostgreSQLContainer<>("postgres:latest")
            .withDatabaseName("football")
            .withUsername("football")
            .withPassword("football");

    static class Initializer
            implements
ApplicationContextInitializer<ConfigurableApplicationContext> {
        public void initialize(ConfigurableApplicationContext
configurableApplicationContext) {
            TestPropertyValues.of(
                    "spring.flyway.url=" + postgreSQLContainer.
getJdbcUrl(),
                    "spring.flyway.user=" + postgreSQLContainer.
getUsername(),
                    "spring.flyway.password=" +
postgreSQLContainer.getPassword(),
                    "spring.r2dbc.url=" + postgreSQLContainer.
getJdbcUrl().replace("jdbc:", "r2dbc:"),
                    "spring.r2dbc.username=" +
postgreSQLContainer.getUsername(),
                    "spring.r2dbc.password=" +
postgreSQLContainer.getPassword())
                    .applyTo(configurableApplicationContext.
getEnvironment());
        }
    }

    @BeforeAll
    public static void startContainer() {
```

```
                    postgreSQLContainer.start();
    }
```

Note that the context configuration gets the database configuration from the PostgreSQL test container. As `PostgreSQLContainer` only returns the JDBC version of the URL, we replaced the `jdbc:` string with `r2dbc:` for the R2DBC driver, while keeping the JDBC URL version for Flyway.

The rest is just standard Flyway configuration. The sample project provides the database initialization scripts in the default folder at `resources/db/migration`.

Event-driven applications with Spring Cloud Stream and RabbitMQ

We want to enhance the experience of our football application by using facts from a match, such as a goal or a red card. We can use the information to prepare a timeline with all events happening during the match or update the match score. We foresee that in the future, we can use this information for other features, such as preparing player statistics in real time or creating a player's ranking using the statistics.

For this scenario, we can apply an **event-driven architecture** design. This type of design consists of detecting, processing, and reacting to real-time events as they happen. Usually, there are two types of components: the event producers and the event consumers, and they are loosely coupled. Spring Cloud Stream is the Spring project that supports event-driven applications that communicate using a shared messaging system, such as Kafka or RabbitMQ.

In this recipe, we'll learn how to use Spring Cloud Stream to create one application that emits football match events and one that is subscribed to those events. We'll also learn how to configure the applications to use RabbitMQ as the messaging system.

Getting ready

In this recipe, we'll use RabbitMQ. You can install it on your computer by following the instructions of the official website at `https://www.rabbitmq.com/docs/download`. I recommend running it on Docker locally. With the following command in your terminal, you can download and run RabbitMQ on Docker:

```
docker run -p 5672:7672 -p 15672:15672 \
 -e RABBITMQ_DEFAULT_USER=packt \
 -e RABBITMQ_DEFAULT_PASS=packt \
 rabbitmq:3-management
```

The image used includes the management portal. You can access it at `http://localhost:15672` using `packt` as username and password.

How to do it...

We'll create two applications and we'll connect them using RabbitMQ. We'll use the power of Spring Cloud Stream to make it possible:

1. We'll start by creating two applications using the Spring Boot Initializr tool:

   - `matches`: This application will produce the match events and will publish them on RabbitMQ

   - `timeline`: This application will process all events published to create a match timeline

   For that, repeat this step for each application. Open `https://start.spring.io` in your browser and use the same options as in *Chapter 1*, in *Creating a RESTful API*, save for the following options:

   - For **Artifact**, type the name of the application, which is `matches` or `timeline`.

   - For **Dependencies**, don't select any starter

   We'll add the following dependency to the three applications:

   ```
   <dependency>
       <groupId>org.springframework.cloud</groupId>
       <artifactId>spring-cloud-stream-binder-rabbit</artifactId>
   </dependency>
   ```

2. The next step will be configuring the `matches` application to emit events:

   I. First, we'll define a record that will be used to represent the match events:

   ```
   public record MatchEvent(Long id, Long matchId,
                            LocalDateTime eventTime, int type,
                            String description, Long player1,
                            Long player2) { }
   ```

   II. You can create a nested builder class to facilitate the manipulation. I'm not including the builder code in this sample for brevity, but you can find an implementation on the book's GitHub repository.

   III. Next, we'll define a bean in the `MatchesApplication` class that configures `Supplier<Message<MatchEvent>>`:

   ```
   @Bean
   public Supplier<MatchEvent> matchEvents() {
       Random random = new Random();
       return () -> {
           return MatchEvent.builder()
                   .withMatchId(1L)
                   .withType(random.nextInt(0, 10))
   ```

```
                    .withEventTime(LocalDateTime.now())
                    .withDescription("random event")
                    .withPlayer1(null)
                    .withPlayer2(null)
                    .build();
        };
    }
```

IV.   This bean generates the `MatchEvent` messages with a random value assigned to the `type` field.

V.   Finally, we'll configure the application to use RabbitMQ. We'll bind the bean we just created to a RabbitMQ exchange named `match-events-topic` and configure the binding to map the *routing key* message property to the `eventType` message header. For that, open the `application.yml` file and set the following configuration:

```
spring:
  rabbitmq:
    host: localhost
    username: packt
    password: packt
    port: 5672
  cloud:
    stream:
      bindings:
        matchEvents-out-0:
          destination: match-events-topic
```

Check that the `spring.rabbitmq` properties are aligned with the parameters used to start the RabbitMQ container in the *Getting ready* section. Verify that the binding name matches the method that exposes the bean.

Now, the `matches` application is ready to start producing events. You can start it now.

3.   Next, we'll configure the `timeline` application. This application will consume all events produced by the `matches` application. For that, do the following:

I.   First, create a `Consumer<MatchEvent>` bean. Open the `TimelineApplication` class and add the following code:

```
@Bean
public Consumer<MatchEvent> processMatchEvent() {
    return value -> {
        System.out.println("Processing MatchEvent: "
                            + value.type());
    };
}
```

II.   For the sake of clarity, we haven't created a library with shared code across the three applications. Therefore, you will need to define the `MatchEvent` record in this project, too.

III. For this project, we only need to configure the RabbitMQ configuration to bind an input queue named `timeline` to the `match-events-topic` exchange created in *Step 2* and the `processMatchEvent` function to the input queue. That can be done by configuring the `application.yml` file as follows:

```yaml
spring:
  rabbitmq:
    host: localhost
    username: packt
    password: packt
    port: 5672
  cloud:
    stream:
      function:
        bindings:
          processMatchEvent-in-0: input
      bindings:
        input:
          destination: match-events-topic
          group: timeline
```

The timeline application is now ready to process messages. Just run the application, and it will start processing messages. You will see the messages when the events are processed in the console.

How it works...

The `spring-cloud-stream-binder-rabbit` dependency includes the Spring Cloud Stream starter, with specific bindings for RabbitMQ. The Spring Cloud Stream starter provides the necessary abstractions to create event-driven applications with no specific references to the underlying messaging technology. In our application, we haven't used any explicit RabbitMQ component in the code. For that reason, we could switch to another messaging system, such as Kafka, just by changing the dependencies and the configuration, but with no code changes.

Spring Cloud Stream allows binding a `Supplier` function registered as a bean to a given destination, in our case, `match-events-topic`. That means that the messages produced by the `Supplier` function are sent to that destination. We use `spring-cloud-stream-binder-rabbit`, which includes the bindings to RabbitMQ. When we start the `matches` application, RabbitMQ creates an **exchange**. An exchange is an intermediary component between the producer and the consumer applications. Producer applications never send messages directly to consumers. Then, depending on the type of exchange, the messages are sent to one or multiple queues or discarded. When we started the matches application and the timeline application was not running yet, the exchange was created in RabbitMQ, but as there were no subscribers, the messages were discarded. In the **RabbitMQ** portal, you would see something like this:

Figure 8.1: match-events-topic in RabbitMQ

As you can see, the application calls the Supplier function once every second, which is the default configuration in Spring Cloud Stream. Still, the messages are discarded as it's not defined yet how to route the messages. Indeed, if you check the **Queues and Streams** tab, you will see that there are no queues defined.

In the timeline application, we configured the binding between the match-event-topic exchange and the destination queue where the messages are forwarded. It's defined in the spring.cloud.stream.binding.input properties. We specified match-events-topic with the destination property, and we defined the target queue with the group property. Then, with the spring.cloud.stream.function.bindings properties, we defined the link between that queue and the function registered as a bean to process the messages. After starting the timeline application, you will see that match-events-topic has a binding that connects the match-events-topic exchange to a queue named match-events-topic.timeline. You can check that in RabbitMQ. It should look like this:

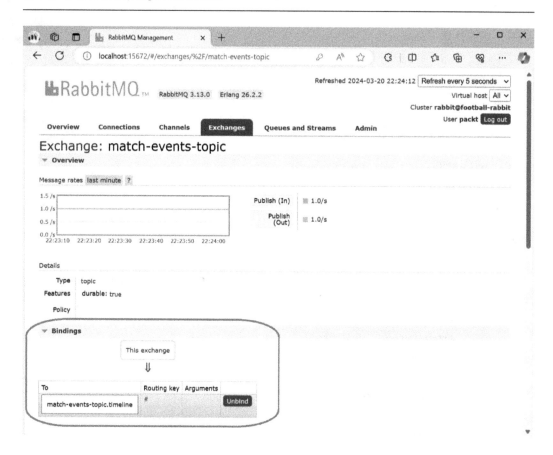

Figure 8.2: Timeline queue bound to match-events-topic

As you can see in the RabbitMQ portal, Spring Cloud Stream created the exchange and the queue and configured the binding to route all messages to the queue.

There's more...

In this recipe, we used the functional approach to send messages, letting the Spring Stream Cloud invoke the `Supplier` function to generate messages. In many scenarios, we must decide when to send messages more explicitly. For this purpose, you can use the `StreamBridge` component. This component lets you send messages while only specifying the binding to use.

Let's see an example. You can create a service component that receives a `StreamBridge` in the constructor, then Spring Boot will inject an instance in runtime:

```
@Service
public class MatchService {
    private StreamBridge streamBridge;
```

```
    private final String bindingName;
    public MatchService(StreamBridge streamBridge,
            @Value("${spring.cloud.stream.bindings.matchEvents-out-0.
destination}") String bindingName) {
        this.streamBridge = streamBridge;
        this.bindingName = bindingName;
    }
}
```

Then, you can use `StreamBridge` to send the message:

```
public void createEvent(MatchEvent matchEvent) {
    streamBridge.send(bindingName, matchEvent);
}
```

See also

RabbitMQ provides different types of exchanges, depending on how the messages are routed:

- **Direct**: The messages are forwarded to a queue based on a routing key.

- **Fanout**: The messages are forwarded to all bounded queues, regardless of the routing key.

- **Topic**: The messages are routed to the bounded queues depending on a pattern defined in the exchange and the routing keys defined in the queues. This is the default type. In the next recipe, we'll explore this scenario further.

- **Headers**: This is similar to the topic exchange, but RabbitMQ routes messages using the message headers instead of the routing key.

Routing messages with Spring Cloud Stream and RabbitMQ

We decided to utilize the match events generated in the previous recipe to update the football match score. We'll create a new application that subscribes to the goal events to implement this functionality.

In this recipe, we'll learn how to configure our Spring Cloud Stream producer application to set the routing key based on the message headers. Then, we'll learn how to configure the consumer applications to set up the queue bindings based on pattern matches for the routing key.

Getting ready

This recipe starts with the outcome of the *Event-driven applications with Spring Cloud Stream and RabbitMQ* recipe. I prepared a working version in case you haven't yet completed that recipe. You can find it in the book's GitHub repository at `https://github.com/PacktPublishing/Spring-Boot-3.0-Cookbook`, in the `chapter8/recipe8-6/start` folder.

As in the *Event-driven applications with Spring Cloud Stream and RabbitMQ* recipe, you will need a RabbitMQ server. To set up the RabbitMQ service on your computer, follow the instructions in the *Getting ready* section of that recipe.

How to do it...

Let's make the few necessary adjustments to the producer application, and then we'll use Spring Cloud Stream's powerful capabilities to set up all RabbitMQ bindings for us:

1. First, we'll modify the `matches` producer application to include some headers in the messages. For that, open the `MatchesApplication` class and modify the `matchEvents` method as follows:

    *   Modify the method's signature; instead of returning `MatchEvent`, we'll return `Message<MatchEvent>`.

    *   We'll use `MessageBuilder` to create the returning `Message<MatchEvent>`.

    *   We'll include a new header named `eventType`. We'll assume that events with the `type` field equal to 2 are goals.

    The method should look like this:

    ```
    @Bean
    public Supplier<Message<MatchEvent>> matchEvents() {
        Random random = new Random();
        return () -> {
            MatchEvent matchEvent = MatchEvent.builder()
                    .withMatchId(1L)
                    .withType(random.nextInt(0, 10))
                    .withEventTime(LocalDateTime.now())
                    .withDescription("random event")
                    .withPlayer1(null)
                    .withPlayer2(null)
                    .build();
            MessageBuilder<MatchEvent> messageBuilder =
                    MessageBuilder.withPayload(matchEvent);
            if (matchEvent.type() == 2) {
                messageBuilder.setHeader("eventType",
    ```

```
                                   "football.goal");
        } else {
            messageBuilder.setHeader("eventType",
                                   "football.event");
        }
        return messageBuilder.build();
    };
}
```

2. Next, we'll change the `matches` application's configuration to assign the `eventType` header's value to the routing key. For that, we'll configure the rabbit producer binding with the `routing-key-expression` property in the `application.yml` file. The Spring Cloud configuration should look like this:

```
spring:
  cloud:
    stream:
      rabbit:
        bindings:
          matchEvents-out-0:
            producer:
              routing-key-expression: headers.eventType
```

I omitted parts already defined in the previous recipe, such as the RabbitMQ host parameters, for readability.

3. Then, we'll create a new consumer application. For that, use the same parameters as in the *Event-driven applications with Spring Cloud Stream and RabbitMQ* recipe; just change the name and use `score`.

4. Next, create a bean that returns `Consumer<MatchEvent>`. You can define it in the `ScoreApplication` class. This function will process the events received. It could look like this:

```
@Bean
public Consumer<MatchEvent> processGoals() {
    return value -> {
        logger.info("Processing goal from player {} at {} ",
                    value.player1(), value.eventTime());
    };
}
```

5. Now, configure the bindings to the `match-event-topic` exchange. In this application, we'll set the binding route key using pattern matching. As we defined in *Steps 1* and *2*, the goals will have the `football.goal` value. So, the configuration will look like this:

```
spring:
  cloud:
    stream:
      rabbit:
        bindings:
          input:
            consumer:
              bindingRoutingKey: football.goal.#
      function:
        bindings:
          processGoals-in-0: input
      bindings:
        input:
          destination: match-events-topic
          group: score
```

6. You can run the `score` application. You should check that it only receives the goal events. If you run the `timeline` application we created in the *Event-driven applications with Spring Cloud Stream and RabbitMQ* recipe, you will see that it receives all events, including the goals.

> **Tip**
> Keep in mind that the Spring Cloud Stream doesn't automatically remove the bindings when you stop the application. For that reason, you may need to manually remove the bindings, exchanges, or queues of previous executions. You can do that from the RabbitMQ portal.

How it works...

The key concept in this recipe is the **routing key**, which is an attribute added by the producer to the message header. The exchange can then use the routing key to decide how to route the message. On the consumer side, it's possible to define a binding to link a queue with an exchange based on the routing keys.

In this recipe, we used the `routing-key-expression` to set the routing key based on the message's properties, such as `header` or `payload`. The Spring RabbitMQ binder allows the use of **Spring Expression Language (SpEL)** to define the message's routing key. The binder evaluates the expression and sets the value of the routing key.

If you look at the messages generated in RabbitMQ, you will see the values of **Routing Key** and **headers**.

Figure 8.3: Message in RabbitMQ showing the routing key

On the consumer side, it's possible to use pattern-matching values when binding the queue to the exchange. The messages whose routing key matches the given pattern will be forwarded to the corresponding queue. If you look at the bindings of match-events-topic in the RabbitMQ portal, you will see that two queues are bound, each using a different routing key.

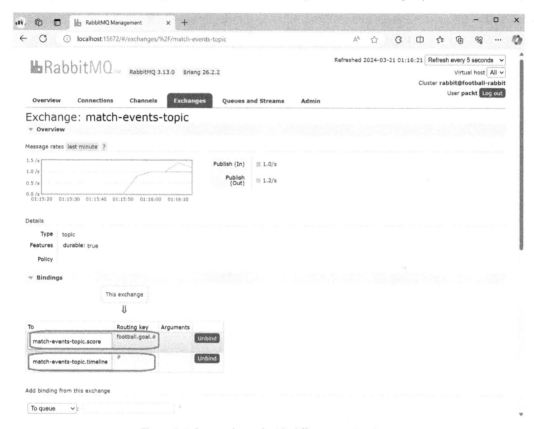

Figure 8.4: Queues bound with different routing keys

With this configuration, RabbitMQ will send to the match-events-topic.score queue only the messages that match the football.goal.# routing key. That is, all messages whose routing key starts with football.goal. It will send to the match-events-topic.timeline queue all messages, as the # symbol acts as a wildcard.

See also

You can find more information about SpEL on the project page at `https://docs.spring.io/spring-framework/reference/core/expressions.html`. It supports querying and manipulating an object graph at runtime, and all Spring projects widely use it.

Error handling with Spring Cloud Stream

The score application we developed in the previous recipe has become critical for football matches as it manages the final match score. For that reason, we need to ensure the solution's robustness.

Spring Cloud Stream can implement highly available, loosely coupled, resilient systems. The reason is that the underlying messaging systems, such as RabbitMQ, provide different mechanisms to ensure the message is delivered to the intended destination. First, the messages can be queued before they are delivered. If, for any reason, the consumer application is not ready yet or has a transient failure, it will process the queued messages once it's ready again. You can use the same mechanism to increase the number of consumers to increase the throughput of your system. You can also configure a retry policy in case of a transient failure in your application or forward it to a special queue, which is named a **dead-letter queue (DLQ)**. The DLQ is a common mechanism in messaging systems; the DLQ can receive all messages that cannot be processed normally for whatever reason; the main purpose is to provide the capability to keep messages that cannot be processed, not only due to technical transient errors but those related to software issues as well.

In this recipe, we'll learn how to configure a DLQ and retry policies for our application, and we'll see how the application behaves in case of errors.

Getting ready

This recipe uses the *Routing messages with Spring Cloud Stream and RabbitMQ* recipe outcomes as the starting point. If you haven't completed it yet, you can use the version I prepared, which can be found in the book's GitHub repository at `https://github.com/PacktPublishing/Spring-Boot-3.0-Cookbook`, in the `chapter8/recipe8-7/start` folder.

As in the *Event-driven applications with Spring Cloud Stream and RabbitMQ* recipe, you will need a RabbitMQ server. To set up the RabbitMQ service on your computer, follow the instructions in the *Getting ready* section of that recipe.

How to do it...

Before we configure the application for error handling, we'll simulate some transient errors in our application:

1. Let's modify our score application to introduce random errors when message processing. We'll simulate an error happening 80% of the time. For that, open the processGoals method of the ScoreApplication class and replace the code with the following:

```
@Bean
public Consumer<MatchEvent> processGoals() {
    Random random = new Random();
    return value -> {
        if (random.nextInt(0, 10) < 8) {
            logger.error("I'm sorry, I'm crashing...");
            throw new RuntimeException("Error processing goal");
        }
        logger.info("Processing a goal from player {} at {} ",
                    value.player1(), value.eventTime());
    };
}
```

2. Now, we'll configure the application to automatically create a DLQ. To do that, open the score application's application.yml and set the autoBindDlq property in the spring.cloud.string.rabbit.bindings.input.consumer section to true. It should look like this:

```
spring:
  cloud:
    stream:
      rabbit:
        bindings:
          input:
            consumer:
              bindingRoutingKey: football.goal.#
              autoBindDlq: true
```

I omitted the rest of the settings for brevity, but you should keep them as they are; you only need to add the highlighted property.

If you run the application now, you'll see a new queue named `match-events-topic.score.dlq`. It's bound to the **dead-letter exchange (DLX)**.

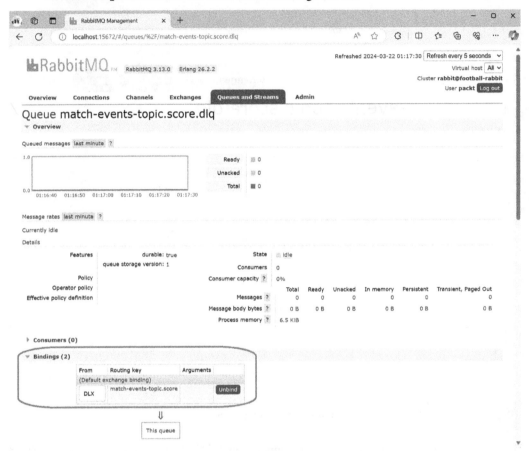

Figure 8.5: DLQ

3. Now, we'll configure the retry policy. For instance, we can configure three attempts before a message is routed to the DLQ and one second as the maximum time to process a message, which is its **time to live (TTL)**. We'll do that by setting the following two properties:

- `spring.cloud.string.rabbit.bindings.input.consumer.maxAttempts = 3`.

- `spring.cloud.string.rabbit.bindings.input.consumer.ttl = 1000`. This property is expressed in milliseconds.

4. You can run the application now. If you keep it running for a while, you will see some errors in the score application, and some messages will reach the DLQ.

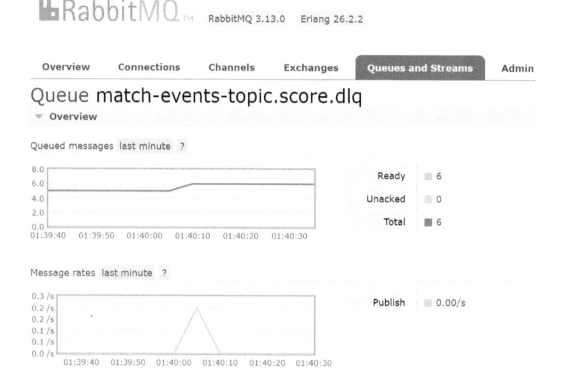

Figure 8.6: DLQ messages

5. You can also stop and restart the `score` application. In the RabbitMQ portal, you can verify that the messages are queued in the `match-events-topic.score` queue and that once the application is ready again, it will process all pending messages.

How it works...

When a queue is bound to an exchange, a new message is forwarded to the queue every time there's a new message in the exchange. The message lasts in the queue until it's processed or the TTL expires. The consumer application, in our case, the `score` application, tries to read and process; if, for any reason, it fails before completing the process, the message returns to the queue. This process can be executed with the number of attempts configured, in our case, three attempts. Finally, if the message cannot be processed or the TTL expires, then the message is forwarded to the DLQ.

In our example, we configured an 80% failure rate for demonstration purposes. If we reduce the failure rate, fewer messages will reach the DLQ.

The solution is more resilient because once the message is queued, we can ensure that it will be processed, regardless of the availability of the `score` application. The `score` application may be unavailable for a given time for many reasons, including a transient error and planned maintenance. For features like these, this kind of solution is so popular in microservices architectures and cloud solutions. The assumption is that the messaging service is highly available, for instance, by deploying a cluster with redundant servers or just by using the **platform as a service (PaaS)** provided by cloud providers. On the Spring Cloud Stream page at `https://spring.io/projects/spring-cloud-stream`, you can find the compatible binders; some are maintained directly by the Spring Cloud Team, while partners maintain others.

See also

You can configure more than one consumer application. If one consumer instance cannot process one message, another instance can process it. With this approach, scaling out the application by adding more consumers is possible. This design is known as **Competing Consumers**. You can find a good description of this scenario in the Azure Architecture Center at `https://learn.microsoft.com/en-us/azure/architecture/patterns/competing-consumers`. The implementation suggests Azure products, but you can apply the same principles with other technologies, such as RabbitMQ.

Other design patterns rely on queue systems that are worth having in our toolbox. For instance:

- **Asynchronous Request-Reply**: In this pattern, you may need to respond quickly to client applications, such as web browsers. However, operations may take much longer to respond. To solve this situation, the application saves the request in a queue and processes it asynchronously. The application exposes an endpoint to get the request status; then, the clients can periodically check the request status.

- **Queue-Based Load Leveling**. Some applications may rely on a backend with limited capacity, but load spikes can occur. In those scenarios with unpredictable demand, the queue acts as a buffer, and the consumer application processes the requests at a pace that doesn't overflow the backend.

Part 4: Upgrading to Spring Boot 3 from Previous Versions

Most of the time invested in an application over its lifetime is related to maintenance. A successful application may last for years or decades. During this time, it may require upgrades for its evolution. You probably have an application that you want to evolve and take advantage of the Spring Boot 3 features. In this part, we'll learn how to upgrade an existing application from Spring Boot 2 to Spring Boot 3.

This part has the following chapter:

- *Chapter 9, Upgrading from Spring Boot 2.x to Spring Boot 3.0*

Upgrading from Spring Boot 2.x to Spring Boot 3.0

Most of the time invested in an application's lifetime is related to maintenance. A successful application may last for years or decades. During this time, it may require upgrades for its evolution. You probably have an application that you want to evolve by taking advantage of the Spring Boot 3 features. In this chapter, we'll use a sample application that I created using Spring Boot 2.6 and perform gradual upgrades in each recipe. The recipes in this chapter should be done in order, as we'll use the outcome of one recipe as the starting point for the next. A couple of recipes won't produce a working version, as there can be compilation errors to be fixed in the following recipes. The last recipe, *Using OpenRewrite for migration automation*, can be done without completing any previous recipes. However, it requires some of the manual actions explained in the previous recipes.

In this chapter, we're going to cover the following recipes:

- Preparing the application
- Preparing Spring Security
- Detecting property changes
- Upgrade the project to Spring Boot 3
- Upgrading Spring Data
- Managing Actuator changes
- Managing web application changes
- Using OpenRewrite for migration automation

Technical requirements

In this chapter, we won't need any additional tools apart from the JDK and the IDE, as in the previous chapters.

Keep in mind that Spring Boot 3.0 requires Java 17 or a later version. To migrate an existing project to Spring Boot 3.0 and use Java 11, you must upgrade to Java 17.

Before migrating to Spring Boot 3.0, I recommend that you upgrade to the latest Spring Boot 2 version, 2.7.x.

In this chapter, we'll use a Spring Boot 2 sample and make changes to it until it's finally migrated to the latest version of Spring Boot 3. This application accesses a PostgreSQL database and a Cassandra database. We'll run both servers using Docker. The official page for running Docker on your computer is https://docs.docker.com/engine/install/. The application has some Testcontainers-based tests, so that you will need Docker on your computer.

To run PostgreSQL on Docker, you can run the following command in your terminal:

```
docker run -itd -e POSTGRES_USER=packt -e POSTGRES_PASSWORD=packt \
-p 5432:5432 --name postgresql postgres
```

You will need to create a database named football. You can do this by running the following PSQL command:

```
CREATE DATABASE football;
```

To run Cassandra on Docker, you can run the following command in your terminal:

```
docker run -p 9042:9042 --name cassandra -d cassandra:latest
```

All the recipes that will be demonstrated in this chapter can be found at: https://github.com/PacktPublishing/Spring-Boot-3.0-Cookbook/tree/main/chapter9.

Next, we will learn about preparing the application.

Preparing the application

On each version of Spring Boot, some components are marked for deprecation, and normally, there is a proposal for change to avoid the deprecated components. As upgrading from Spring Boot 2 to Spring Boot 3 is a major change, upgrading to the latest Spring Boot 2 version is strongly recommended.

Before migrating an application to Spring Boot 3, the following preparation is recommended:

- Upgrade the Spring Boot version to the latest 2.7.x available. At the time of writing this book, it is 2.7.18. This will facilitate the upgrade to Spring Boot 3

- Update the Java version to Java 17, the minimum supported version in Spring Boot 3.

- Address all deprecated components.

In this recipe, we'll prepare a sample application that uses Spring 2.6 and Java 11. By the end of the recipe, the application will use Spring 2.7.18, Java 17, and all deprecated components and APIs will be addressed.

Getting ready

You will prepare an application using Spring 2.6 and Java 11 in this recipe. The sample application is in the book's GitHub repository at `https://github.com/PacktPublishing/Spring-Boot-3.0-Cookbook` in the `chapter9/football` folder.

Depending on your operating system, you may use different tools to manage the current JDK version. For instance, you can use SDKMAN! tool on Linux and Mac. You can install it by following the instructions on the project page at `https://sdkman.io/`.

To run the application, you will need a PostgreSQL server running on your computer. To acquire this, follow the instructions in the *Technical requirements* section.

How to do it...

Let's put the application at the starting line. Ready, steady, go!

1. First, we'll address all deprecation warnings. For instance, if you compile the football project by executing `mvn compile`, you will see one deprecation warning about the `DataSourceInitializationMode` class. If you open the documentation of that class at `https://docs.spring.io/spring-boot/docs/2.6.15/api/org/springframework/boot/jdbc/DataSourceInitializationMode.html`, you will see the following information:

   Deprecated.

   since 2.6.0 for removal in 3.0.0 in favor of DatabaseInitializationMode

You can see this information directly in the editor if you use an IDE, such as IntelliJ or Visual Studio Code. For instance, in Visual Studio Code, you would see the following:

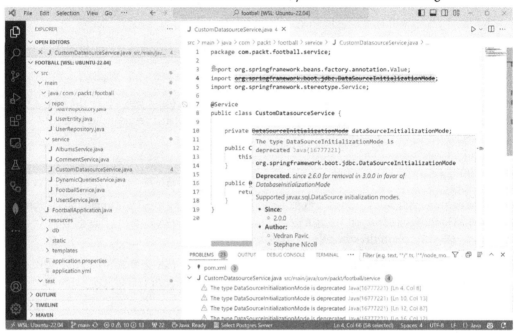

Figure 9.1: Deprecation message in Visual Studio Code

2. Now, let's replace DataSourceInitializationMode with DatabaseInitializationMode. This change is very straightforward, as just a simple replacement is enough. Other deprecation changes may require some code refactoring. Usually, the deprecated class documentation guides the implementation of the changes.

3. The next step will be updating the Spring Boot version to 2.7.18. For that, open the pom.xml file and find the following code snippet:

```
<parent>
    <groupId>org.springframework.boot</groupId>
    <artifactId>spring-boot-starter-parent</artifactId>
    <version>2.6.15</version>
    <relativePath/> <!-- lookup parent from repository -->
</parent>
```

That refers to the Spring Boot version in use. In the highlighted code, you can see that the actual version is 2.6.15. Let's update it to the latest version 2 available. At the time of writing this book, it is 2.7.18.

If you recompile the application, you will see some deprecation warnings related to Spring Security in the `SecurityConfiguration.java` file. We'll fix those warnings in the *Upgrading Spring Security* recipe.

4. Now, update the Java version to Java 17. You should ensure you are using the JDK 17:

   - If you are on Linux or Mac, you can use SDKMan!

   - On Windows, you may have to update the Java Home environment variable.

   - Regardless of the operating system, you can manage the Java version using the IDE:

     - For instance, in Visual Studio Code, you can follow the instructions at `https://code.visualstudio.com/docs/java/java-project#_configure-runtime-for-projects` to configure the Java runtime for your project.

     - In IntelliJ, you can follow the instructions at `https://www.jetbrains.com/help/idea/sdk.html#manage_sdks` for this purpose.

   You should change the Java version in your project as well. For that, open the `pom.xml` file and look for the property `java.version`:

   ```
   <properties>
       <java.version>11</java.version>
       <testcontainers.version>1.19.7</testcontainers.version>
   </properties>
   ```

   Replace the `java.version` property with `17`.

5. Build the application again and verify that the only deprecation warnings are related to Spring Security in the `SecurityConfiguration.java` file. Again, if there are other deprecation warnings, try to use the alternative solution proposed by the deprecation warning message.

How it works...

Every time there is a new version, there can be components marked for deprecation, and as we saw in this recipe, usually, it's also known in which version they will be removed from. It's convenient to install the upgrades gradually and not skip intermediate versions. As we saw in this example, migrating all revision versions, such as 2.7.1, 2.7.2, and so on, is unnecessary. However, you should not skip minor versions, such as 2.7.x, as there can be new components marked for deprecation that will be removed in version 3. Skipping the 2.7 upgrade won't let you see the warning and the alternative replacement. Suddenly, you will find that the class you used is not found.

See also

Spring Boot 3.0 uses Spring Framework 6.0. If your project has explicit dependencies on prior versions, you will also need to upgrade them. You can find the migration guide for Spring Framework at `https://github.com/spring-projects/spring-framework/wiki/Upgrading-to-Spring-Framework-6.x`.

Preparing Spring Security

The main change regarding security in Spring Boot 3 is the upgrading from Spring Security 5 to Spring Security 6. There are many changes related to this upgrade, but in the recipe, we'll focus on the most common one, which is how it's configured.

In Spring Security 5, most of the settings were configured by extending the `WebSecurityConfigurerAdapter` class, and in Spring Security 6, those changes are applied by configuring specific beans in our application.

In this recipe, we'll transform the `WebSecurityConfigurerAdapter` class into a configuration class that exposes the beans to apply an equivalent configuration.

Getting ready

The starting point of this recipe is the outcome of the *Preparing the application* recipe. I prepared a working version in case you haven't completed it yet. You can find it in the book's GitHub repository at `https://github.com/PacktPublishing/Spring-Boot-3.0-Cookbook` in the `chapter9/recipe9-2/start` folder.

How to do it...

Let's complete the preparation of our application by adapting the deprecated components related to Spring Security:

1. If you compile the project as-is now, you will see that the `SecurityConfig` class contains some deprecation warnings. Let's tackle them:

    -   The `WebSecurityConfigurerAdapter` class is deprecated. Now, the application won't extend any class.

    -   The `withDefaultPasswordEncoder` method is deprecated and not recommended for production environments. However, as the documentation states, it's acceptable for demos and getting started. We won't change it.

2. The `SecurityConfig` class is annotated with `@EnableWebSecurity`. In addition, it should be annotated with `@Configuration`. It should look like this:

```
@Configuration
@EnableWebSecurity
public class SecurityConfig
```

3. Next, we'll change the `configure` method that receives `AuthenticationManagerBuilder` as a parameter to create a method configuring an `InMemoryUserDetailsManager` bean:

```
@Bean
InMemoryUserDetailsManager userDetailsManager() {
    UserDetails userAdmin = User.withDefaultPasswordEncoder()
            .username("packt")
            .password("packt")
            .roles("ADMIN")
            .build();
    UserDetails simpleUser = User.withDefaultPasswordEncoder()
            .username("user1")
            .password("user1")
            .roles("USER")
            .build();
    return new InMemoryUserDetailsManager(userAdmin,
simpleUser);
}
```

4. Now, we'll replace the `configure` method that receives a `WebSecurity` with a method returning a `WebSecurityCustomizer` bean. It should look like this:

```
@Bean
WebSecurityCustomizer webSecurityCustomizer() {
    return (web) -> web.ignoring()
            .antMatchers("/security/public/**");
}
```

5. To complete the `SecurityConfig` class, we must replace the `configure` method receiving a `HttpSecurity` parameter with a method creating a `SecurityFilterChain` bean:

```
@Bean
SecurityFilterChain filterChain(HttpSecurity http)
                                            throws Exception {
    return http
            .authorizeRequests(authorizeRequests -> {
                try {
                    authorizeRequests
                            .antMatchers("/").permitAll()
```

```
                            .antMatchers("/security/private/**")
                                .hasRole("ADMIN")
                            .and()
                            .httpBasic();
                } catch (Exception e) {
                    e.printStackTrace();
                }
            }).build();
    }
```

6. Now, if we execute the tests, we'll realize that the tests related to controllers using the @WebMvcTest annotation no longer work as expected. We'll include the @Import annotation in the FootballControllerTest and SecurityControllerTest classes to fix it.

    • FootballControllerTest:

    ```
    @WebMvcTest(FootballController.class)
    @Import(SecurityConfig.class)
    class FootballControllerTest
    ```

    • SecurityControllerTest:

    ```
    @WebMvcTest(SecurityController.class)
    @Import(SecurityConfig.class)
    class SecurityControllerTest
    ```

7. Finally, execute the tests and verify that all of them pass.

How it works...

The main change is the deprecation of the WebSecurityConfigurerAdapter class. This class is removed in Spring Security 6 and, hence, from Spring Boot 3. In this recipe, we prepared the application for a smooth security migration to upgrade the project to Spring Boot 3. The new approach is to create beans for each security aspect we want to configure.

Spring Security 6 introduces new methods to replace the antMatcher method; for instance, the requestMatcher method. To avoid more changes in this stage, we haven't replaced the antMatcher method yet. But we'll do that once we upgrade to Spring Boot 3, as it will be deprecated.

We keep using withDefaultPasswordEncoder in this recipe. Although it's deprecated and not recommended for production environments, using it in development environments is acceptable. The deprecation warning is present in Spring Boot 2, and this method won't be removed any time soon. Deprecation was introduced to warn about usage in production environments.

As the security settings are now defined as a configuration class annotated with `@Configuration`, importing them into our `@WebMvcTest` tests is necessary. The `@WebMvcTest` annotation only loads the MVC-related components, and it doesn't load `SecurityConfig` by default. For that reason, importing them into our `@WebMvcTest` tests is necessary.

See also

The upgrade from Spring Security 5 to Spring Security 6 includes more changes than the ones tackled in this recipe. You can find the full migration guide on the official website at `https://docs.spring.io/spring-security/reference/6.0/migration/index.html`.

Detecting property changes

As we learned in this book, one of the main tasks when developing a Spring Boot application is configuring its components. It's important to note that every time a new component version is released, there may be changes to the properties. This is especially true when upgrading to a major version, such as from version 2.x to 3.0. Typically, the changes are gradual. For example, a property might be marked as *deprecated* in one version, meaning it will be removed in a future version. Therefore, it's recommended not to skip any versions when upgrading. For instance, if you plan to upgrade from version 2.6 to version 3.2, it's best to upgrade first to version 2.7, then to 3.0, then 3.1, and finally 3.2.

To address the property changes between versions, Spring Boot provides a tool named Spring Boot Properties Migrator. It's a dependency that analyzes the application environment during the application start-up and prints a diagnostic.

In this recipe, we'll use the Spring Boot Properties Migrator in our project to detect the deprecated properties in the configuration file and fix them.

Getting ready

We'll start this recipe using the outcome of the *Preparing Spring Security* recipe. You can use the project I prepared as a starting point if you haven't completed it yet. You can find it in the book's GitHub repository at `https://github.com/PacktPublishing/Spring-Boot-3.0-Cookbook` in the `chapter9/recipe9-3/start` folder.

You will need the PostgreSQL server and Cassandra running on your computer to run the application. For instructions, see the *Technical requirements* section in this chapter.

How to do it...

Our application uses deprecated properties; let's fix them with Spring Boot Properties Migrator!

1. First, we'll add the Spring Boot Properties Migrator dependency to our project. For that, open the pom.xml file and add the following dependency:

```
<dependency>
    <groupId>org.springframework.boot</groupId>
    <artifactId>spring-boot-properties-migrator</artifactId>
    <scope>runtime</scope>
</dependency>
```

2. Next, you can run the application and see an error with the property in the properties file that should be migrated. For instance, I executed mvn spring-boot:run in the terminal and I received the following message:

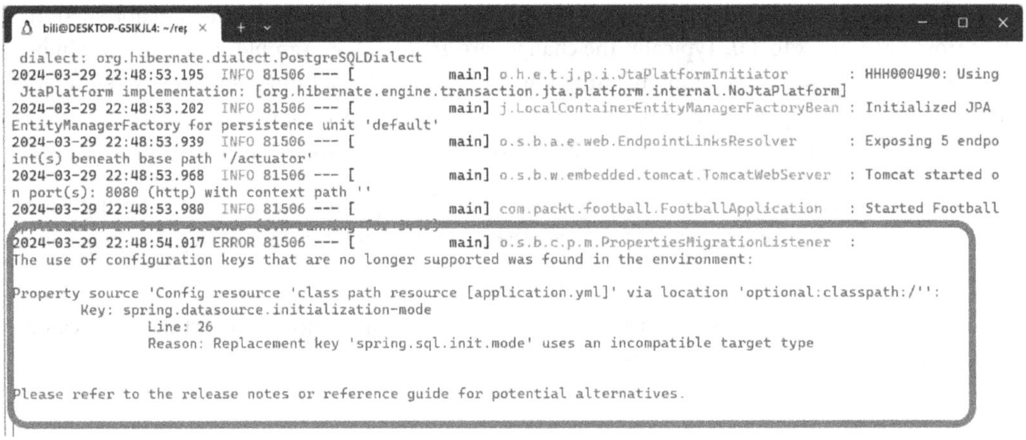

Figure 9.2 – Error when executing the application with Spring Boot Properties Migration

As you can see, we need to migrate the spring.datasource.initialization-mode property and use spring.sql.init.mode instead.

3. Then, we'll replace the spring.datasource.initialization-mode property with spring.sql.init.mode. To do this, open the application.yml file, find the initialization-mode property, and remove it. Then, add the spring.sql.init.mode property. The spring configuration should look like this:

```
spring:
    application:
        name: football
    jpa:
        database-platform: org.hibernate.dialect.
```

```
PostgreSQLDialect
        open-in-view: false
    datasource:
        url: jdbc:postgresql://localhost:5432/football
        username: packt
        password: packt
        hikari:
            maximum-pool-size: 4
  sql:
    init:
      mode: embedded
```

For clarity, I didn't include the entire `application.yml` file. The complete file can be found in the book's GitHub repository.

4. The `dataSourceInicitalizationMode` parameter of the `CustomDatasourceService` class references the `spring.datasource.initialization-mode` configuration property. We must point to the new property, `spring.sql.init.mode`. For that, replace the `@Value` annotation in the constructor with the new property. It should look like this:

```
public CustomDatasourceService(@Value("${spring.sql.init.mode}")
DatabaseInitializationMode dataSourceInitializationMode) {
    this.dataSourceInitializationMode =
dataSourceInitializationMode;
}
```

5. Now that the changes are applied, let's execute the application again. You can do it by running the `spring-boot:run` command in your terminal.

You will see that the application runs normally.

6. Now that the properties have been migrated, the last step is to remove the Spring Boot Properties Migration dependency. To do this, open the `pom.xml` file and remove the dependency you added in *step 1*.

How it works...

The Spring Boot Properties Migration dependency first tries to map the deprecated properties to the new ones when possible and then prints a warning. Without direct mapping to a new property, it throws an error and prints the issue. In this recipe, we used a dependency that Spring Boot Properties Migration cannot map automatically. For that reason, there was an error. Once fixed, the Spring Boot Migration doesn't show any additional errors.

As the Spring Boot Properties Migration performs an additional check upon application start-up, it can slow down the application. Removing the Spring Boot Properties Migration from your application is a good practice once the deprecated properties are fixed.

Upgrade the project to Spring Boot 3

In this recipe, we'll take the first steps in Spring Boot 3 by updating the references in our project. When we update to Spring Boot 3, we'll see some compilation errors that need to be addressed; in this recipe, we'll take the first steps toward fixing them.

Spring Boot 3 depends on Jakarta EE 9 or a later version, while Spring Boot relies on Jakarta 7 and 8. In Jakarta 9, all namespaces changed from `javax.*` to `jakarta.*`. This is probably the biggest impact that can be seen when upgrading to Spring Boot 3, as it requires changing many references in our projects.

In this recipe, we'll finally upgrade our project to Spring Boot 3, which will require updating all `javax` namespace references to `jakarta`. We'll also perform the latest updates to Spring Security. By the end of this recipe, the application won't work yet; you must complete two recipes, *Upgrading Spring Data* and *Managing Actuator changes*, to make it work.

Getting ready

We'll use the outcome of the *Detecting property changes* recipe as the starting point for this recipe. You can use the version I prepared if you haven't completed it yet. You can find it in the book's GitHub repository at `https://github.com/PacktPublishing/Spring-Boot-3.0-Cookbook` in the `chapter9/recipe9-4/start` folder.

How to do it...

We'll upgrade the application to Spring Boot 3, and we'll see the Jakarta EE-related errors. Let's fix them and start using Spring Boot 3!

1. First, we'll upgrade the project to use Spring Boot 3. For that, open the `pom.xml` file and find the following code snippet:

   ```
   <parent>
       <groupId>org.springframework.boot</groupId>
       <artifactId>spring-boot-starter-parent</artifactId>
       <version>2.7.18</version>
       <relativePath/> <!-- lookup parent from repository -->
   </parent>
   ```

   You should replace the version property with the latest Spring Boot version, which, at the time of writing this book, is 3.2.4.

2. You will see that the project doesn't compile, as all references to `javax.*` namespaces no longer work. To fix it, replace `javax` with `jakarta` in all files. There are a significant number of changes to make, so I recommend using the replace features of your favorite editor. For instance, you can use the **Search: Replace in files** command in Visual Studio Code:

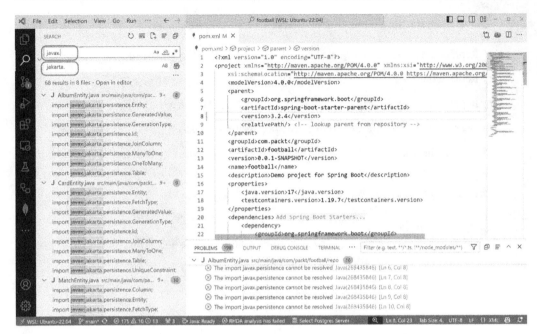

Figure 9.3 : Replacing javax with jakarta in files in Visual Studio Code

3. If you try to compile the project, you will see that there are still three files with compilation errors:

    • `MatchEventEntity.java`: There are errors related to fields mapped as JSON in the database. This issue will be tackled in the *Upgrading Spring Data* recipe.

    • `FootballConfig.java`: There are errors related to `HttpTrace`. This issue will be fixed in the *Managing Actuator changes* recipe.

    • `SecurityConfig.java`: There are new errors related to Spring Security changes that we'll fix in the following step.

4. To fix the issues with the `SecurityConfig` class, we need to do the following:

    • Replace the `antMatchers` call with a `requestMatchers` call in the `webSecurityCustomizer` method. It should look like this:

```
@Bean
WebSecurityCustomizer webSecurityCustomizer() {
    return (web) -> web.ignoring()
            .requestMatchers("/security/public/**");
}
```

- In the `filterChain` method, we need to make several changes:

  - Replace `authorizeRequests` with `authorizeHttpRequests`.

  - Replace the `requestMatchers("/")` with `anyRequest()`.

  - Replace the `antMatchers` call with a `requestsMatchers`.

  - We need to switch the call order in the previous two calls.

  - Remove the `and` call and just chain `authorizeHttpRequests` with the `httpBasic` call.

```
@Bean
SecurityFilterChain filterChain(HttpSecurity http)
                                            throws Exception {
    return http
            .authorizeHttpRequests(authorizeRequests -> {
                authorizeRequests
                        .requestMatchers("/security/private/**")
                            .hasRole("ADMIN")
                        .anyRequest().permitAll();
            })
            .httpBasic(withDefaults())
            .build();
}
```

How it works...

Jakarta EE was formerly **Java EE**, which stands for **Java Enterprise Edition**. Java EE is a set of API specifications for building enterprise-grade applications. You can build your application based on these specifications and select the vendor implementation that fits your requirements without changing your code or with little changes. In 2017, Oracle transferred Java EE to the Eclipse Foundation and then changed its name to Jakarta EE. With Jakarta EE version 9, the package names shifted from `javax` to `jakarta`, allowing Jakarta EE to evolve independently from Oracle, as it keeps some Java trademarks. These changes only imply namespace changes, and there are no behavior changes. For that reason, the application works just by replacing the namespaces.

In this recipe, we moved from Spring Boot 2.7.18 directly to Spring Boot 3.2.4. If we first moved to intermediate versions, we would see the Spring Security-related errors as deprecation warnings. As we know the mechanism for gradual deprecation and upgrades, I decided to skip the intermediate versions for brevity. I recommend that you perform gradual migrations in your projects. The deprecation warning messages can guide most of the changes. However, there's a change that's not very evident. In Spring Boot 6, the `permitAll` method is deprecated for `requestMatcher` and should be replaced by `anyRequest`, and this method should be called at the end of the chain; if you chain more matchers after, an exception is thrown during runtime.

Upgrading Spring Data

Spring Boot 3 uses Hibernate 6.1 by default, while Spring Boot 2 uses Hibernate 5. Therefore, we need to prepare the application to match Hibernate 6.1.

Hibernate uses Jakarta EE, and that requires upgrading the `javax.*` namespaces to `jakarta.*`, but we already did this step in the *First step to Spring 3.0* recipe.

Some changes in Hibernate 6.1 are internal, but some APIs changed and should be upgraded in the application.

Some changes related to the configuration of Spring Data are specific to Cassandra.

In this recipe, we'll make the necessary changes to align the sample application with Hibernate 6.1 and Cassandra.

Getting ready

We'll use the outcome of the *First step to Spring 3.0* recipe as the starting point for this recipe. You can use the version I created if you haven't completed it yet. You can find it in the book's GitHub repository at `https://github.com/PacktPublishing/Spring-Boot-3.0-Cookbook` in the `chapter9/recipe9-5/start` folder.

As explained in the previous recipes, the sample application uses a PostgreSQL server database and Cassandra, and the tests use Testcontainers. For that reason, you need Docker running on your computer.

How to do it...

Let's make the necessary Spring Data adjustments in our application:

1. We'll start by fixing the configuration of our application:

    *   We'll replace `PostgreSQL82Dialect` with `PostgreSQLDialect`. To do that, we'll open the `application.yml` file, locate the `spring.jpa.database-platform` property, and then set the following value:

        ```
        spring:
          jpa:
            database-platform: org.hibernate.dialect.PostgreSQLDialect
        ```

    *   We'll fix the Cassandra settings: `spring.data.cassandra.*` settings now should be `spring.cassandra.*`. In the same `application.yml` file, ensure that the cassandra settings are defined as follows:

        ```
        spring:
          cassandra:
              keyspace-name: footballKeyspace
        ```

```
schema-action: CREATE_IF_NOT_EXISTS
contact-points: localhost
local-datacenter: datacenter1
session-name: cassandraSession
port: 9042
```

Note that `cassandra` is now under `spring`. Previously, `cassandra` was under `spring. data`.

2. The tests based on Testcontainers create a Cassandra container and then set the settings of the application context. The tests should be aligned with the new settings structure with `spring. cassandra` instead of `spring.data.cassandra`. You can apply this change by using the string replace features in your editor. For instance, you can use the **Search: Replace in files** feature in Visual Studio Code:

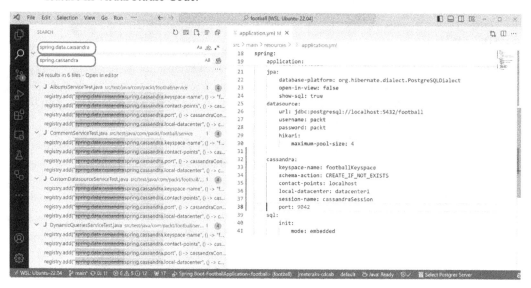

Figure 9.4 : Replacing spring.data.cassandra references in Visual Studio Code

3. Next, we'll use the new capability in Hibernate 6 to define JSON fields for our entities. To do that, open the `MatchEventEntity` class and modify the annotation of the `details` field as follows:

    ```
    @JdbcTypeCode(SqlTypes.JSON)
    private MatchEventDetails details;
    ```

    We replaced the *hypersistence utils* type to define the JSON field.

4. As we no longer need the particular type to define the JSON field, you can remove the following dependency from the pom.xml file:

```
<dependency>
    <groupId>io.hypersistence</groupId>
    <artifactId>hypersistence-utils-hibernate-55</artifactId>
    <version>3.7.3</version>
</dependency>
```

5. At this point, we want to leverage the project's tests. However, there are some compilation errors related to the httptrace endpoint. We'll now comment all the code in the FootballConfig class to avoid the compilation errors, and we'll tackle this component in the *Managing Actuator changes* recipe.

6. Next, we'll run the tests to verify that the application still works. You can run the tests from the IDE or just run the following Maven command in your terminal:

```
mvn test
```

You will see that the tests related to the services succeed, but two tests, findPlayerById and countPlayers, fail. There are other tests related to the controllers that fail, but we'll cover them in the *Managing web application changes* recipe. These tests fail due to some behavior changes in Hibernate 6. Let's fix them:

- The findPlayerById method in the DynamicQueriesService class uses an ordinal parameter binding. The behavior of ordinal parameters was changed in Hibernate 6. To fix this issue, modify findPlayerById in the DynamicQueriesService class as follows:

```
public Player findPlayerById(Integer id) {
    Query query = em.createQuery(
                    "SELECT p FROM PlayerEntity p WHERE
p.id=?1",
                    PlayerEntity.class);
    query.setParameter(1, id);
    return playerMapper.map((PlayerEntity)
                                query.getSingleResult());
}
```

- The change is subtle. In the query string, replace ?0 with ?1; the parameter in the setParameter method is 1 instead of 0.

- The countPlayers test validates the method with the same name in the DynamicQueriesService class. To fix this test, do the following:

    i.   First, change the return type of the countPlayers method in the DynamicQueriesService class from BigInteger to Long. It should look like this:

```
public Player findPlayerById(Integer id) {
    Query query = em.createQuery(
            "SELECT p FROM PlayerEntity p WHERE p.id=?1",
            PlayerEntity.class);
    query.setParameter(1, id);
    return playerMapper
            .map((PlayerEntity) query.getSingleResult());
}
```

    ii.   Then, update the test to match the return type with Long:

```
@Test
void countPlayers() {
    Long count = dynamicQueriesService.countPlayers();
    assertThat(count, not(0));
}
```

You can rerun the services-related tests, and now they should succeed.

How it works...

Spring Data relies on Hibernate for most of its functionalities. In Spring Boot 3, it uses Hibernate 6.1 by default. For that reason, most of the tasks related to the Spring Data upgrade are related to the Hibernate upgrade.

One of the changes related to the Hibernate upgrade is the change in the references from javax.* to jakarta.*. We haven't explained that in this recipe, as it was already covered in the *First step to Spring 3.0* recipe; you should keep it in mind when upgrading to Spring Boot 3 or Hibernate 6.

In Hibernate 6, the property spring.jpa.database-platform no longer uses specific version values. For that reason, PostgreSQL82Dialect was deprecated and should be replaced with the database dialect without the version. As we use PostgreSQL as a relational database, we replaced PostgreSQL82Dialect with PostgreSQLDialect. If you use another database engine such as MySQL, you should use MySQLDialect without a version-specific dialect.

In Hibernate 6, the queries that return a BIGINT are now mapped to the Long type. In previous versions, they were incorrectly mapped to BigInteger. The result of the count clause is a BIGINT, so we need to change it in the countPlayers method. Even though the number of players in our application can be represented by an integer, if our tables were larger, it could cause a casting error in runtime.

Hibernate 6 changed how it binds the ordinal parameters and now uses 1-based ordering instead of 0-based ordering.

Hibernate 6 introduced the possibility of defining entity fields mapped as JSON parameters if the database supports it. Before Hibernate 6, it was necessary to use third-party libraries, such as Hypersistence. This excellent library developed by Vlad Mihalcea was handy in the previous versions for JSON field management, but it's no longer necessary in Hibernate 6. We can keep this dependency and upgrade it to the matching version for Hibernate 6. You can find more information at `https://github.com/vladmihalcea/hypersistence-utils`.

There are other changes related to Spring Data that are not related to Hibernate. For instance, the `spring.data` prefix for the properties is now reserved for Spring Data; for that reason, `spring.data.cassandra` has moved to `spring.cassandra`.

In this recipe, we only covered the changes related to Spring Data for the features we used in this book. I recommend that you check the migration guide at `https://github.com/spring-projects/spring-boot/wiki/Spring-Boot-3.0-Migration-Guide#data-access-changes`.

There's more...

Spring Boot 3 introduced some capabilities that facilitate the usage of standard database technologies in our applications. Before Spring Boot 3, we could use alternative solutions, and it's unnecessary to migrate them; however, it is worth knowing they exist. In this section, we'll demonstrate two of them: stored procedures and union clauses in JQL.

You can use the `@Procedure` annotation instead of the `@Query` annotation. The `@Query` annotation must be native and use the `call` clause. For instance, you can modify the `getTotalPlayersWithMoreThanNMatches` method of the `PlayerRepository` interface as follows:

```
@Procedure("FIND_PLAYERS_WITH_MORE_THAN_N_MATCHES")
Integer getTotalPlayersWithMoreThanNMatches(int num_matches);
```

Hibernate 6 JQL now supports the Union clause. For instance, we can write `findPlayersByMatchId` in the `MatchRepository` interface as follows:

```
@Query("SELECT p1 FROM MatchEntity m JOIN m.team1 t1 JOIN t1.players
p1 WHERE m.id = ?1 UNION SELECT p2 FROM MatchEntity m JOIN m.team2 t2
JOIN t2.players p2 WHERE m.id = ?1")
public List<PlayerEntity> findPlayersByMatchId(Integer matchId);
```

See also

In this recipe, we covered some scenarios on the Hibernate 5 to 6 migration, but there are many more. If you find issues related to Hibernate in the migration of your projects, please take a look at the Hibernate migration guides:

- Hibernate 6.0 migration guide: `https://docs.jboss.org/hibernate/orm/6.0/migration-guide/migration-guide.html`

- Hibernate 6.1 migration guide: `https://docs.jboss.org/hibernate/orm/6.1/migration-guide/migration-guide.html`

Managing Actuator changes

Most of the Actuator changes are related to the default behavior of the exposed endpoints. For instance, the JMX endpoint behavior is aligned with the web endpoint behavior; for that reason, it only exposes the **Health** endpoint by default, while previously, all JMX endpoints were exposed by default. If your project relies on that functionality, you must expose it explicitly.

In addition to Actuator's default behavior, our project uses the `httptrace` endpoint, which changes the behavior and the required implementation. In this recipe, we'll fix the `httptrace` endpoint and make the necessary configuration changes to keep the same behavior we had for Actuator.

Getting ready

To start this recipe, you will need the outcome of the previous recipe, *Upgrading Spring Data*. In case you haven't completed it yet, I prepared a version that you can find in the book's GitHub repository at `https://github.com/PacktPublishing/Spring-Boot-3.0-Cookbook` in the `chapter9/recipe9-6/start` folder.

In addition to the requirements of our previous recipes, we'll need a JMX client to explore the behavior of the Actuator's JMX endpoints. We can use JConsole, which is part of the JDK.

How to do it...

We'll first fix the `httptrace` endpoint, and then we'll align the actuator to behave as in Spring Boot 2:

1. To fix the `httptrace` configuration, open the `application.yml` file and replace the following:

   - Replace the web endpoint `httptrace` with `httpexchanges`.

   - Replace the `management.trace.http.enabled` with `management.httpexchanges.recording.enabled`.

- It should look like this:

```
management:
  endpoints:
    web:
      exposure:
        include: health,env,metrics,beans,loggers,httpexchanges
  httpexchanges:
    recording:
      enabled: true
```

For clarity, I am not including the rest of the `application.yml` file.

2. Next, for the `FootballConfig` class, replace the `HttpTraceRepository` interface with `HttpExchangeRepository` and `InMemoryHttpTraceRepository` with `InMemoryHttpExchangeRepository`. Remember that we commented the contents of this class in the previous recipe to be able to compile the solution; now, we'll tackle this component. The `FootballConfig` class should look as follows:

```
@Configuration
public class FootballConfig {
    @Bean
    public HttpExchangeRepository httpTraceRepository() {
        return new InMemoryHttpExchangeRepository();
    }
}
```

3. We can now run the application and validate the new `httpexchanges` repository. To validate it, open it in your browser or execute a curl command with the address `http://localhost:8080/actuator/httpexchanges`. It should return the latest requests to our application.

4. Next, we'll validate whether the application exposes the same JMX endpoints as Spring Boot 2. While the application is running, run the JConsole tool. For that, open your terminal and run `jconsole`. You will see that it shows a list of the Java applications running on your computer:

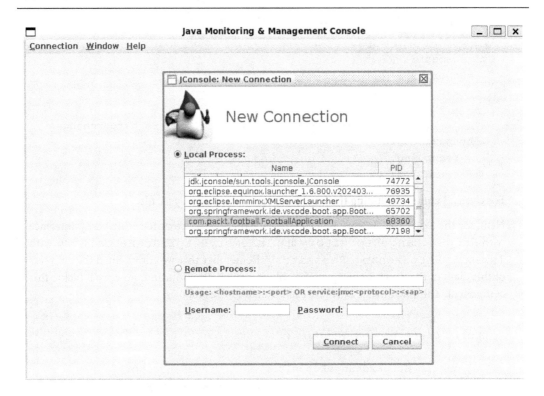

Figure 9.5: JConsole process selection

Select the application and click **Connect**. A message indicating that the secure connection failed may appear, suggesting that you connect insecurely. Use the insecure connection.

If you open the **MBeans** tab, you'll see that under the `org.springframework.boot` namespace, only the **Health** endpoint appears:

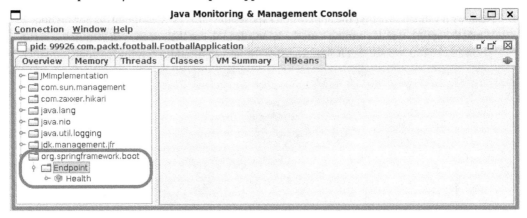

Figure 9.6: Default JMX endpoints exposed by Spring Boot 3

Before Spring Boot 3, all endpoints were enabled by default. To achieve the same behavior, open the `application.yml` file and set `management.endpoints.jmx.exposure.include=*` property. It should look as follows:

```
management:
    endpoints:
        jmx:
            exposure:
                include: '*'
```

If you restart the application and connect to the application with JConsole, you will see that it now exposes all MBean endpoints, as in previous versions of Spring Boot:

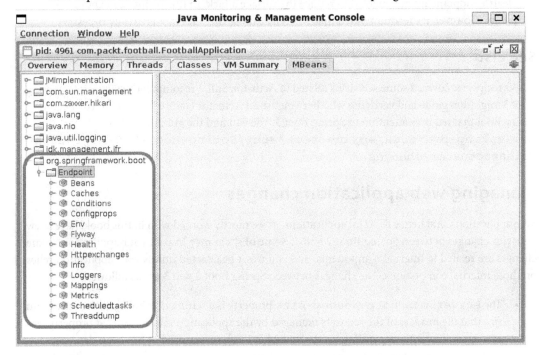

Figure 9.7: All JMX endpoints exposed using Spring Boot 3

You can see the exposed endpoints in *Figure 9.7*.

How it works...

The Spring Boot team changed the name of httptrace to avoid confusion with Micrometer Tracing. For that reason, it has been renamed to http exchanges. That change also impacted the repository supporting the implementation of the traces. In this example, we used an in-memory repository, which we used only for demonstration purposes. In a production environment, you will probably use a persistent repository.

In Spring Boot 3, the JMX endpoint only exposes the **Health** endpoint to align with its web counterpart. Only the **Health** endpoint is exposed in the web endpoint in Spring Boot 3 and previous versions. Some endpoints can reveal sensitive information or provide unwanted access. All endpoints except the **Health** endpoint are not enabled to reduce the surface attack. In this recipe, we exposed all JMX endpoints. However, it's recommended that you only expose the ones that are really necessary.

See also

In this recipe, we covered some scenarios related to Actuator. Still, I recommend reviewing the Spring Boot 3 migration guide and verifying whether you use an Actuator feature from previous versions of Spring Boot that requires attention in Spring Boot 3. You can find the guide at https://github. com/spring-projects/spring-boot/wiki/Spring-Boot-3.0-Migration-Guide#actuator-changes.

Managing web application changes

Web applications, and hence RESTful applications, as we mostly worked with in this book, have a few behavior changes between Spring Boot 2 and 3. Some of them may impact your application. Some changes are related to internal components, and you won't be affected unless your application relies on those internal components. The changes between Spring Boot 2 and 3 are as follows:

- The server.max-http-header-size property is a setting that indicates the maximum size that the headers of the requests managed by the application may have. That property was managed differently depending on the embedded web server used. It has been deprecated in favor of server.max-http-request-header-size, and it's managed consistently by all possible embedded web servers.

- The phases for graceful shutdown have changed. When an application is shut down, Spring Boot sends events in different phases, and the application can be subscribed to those events to perform custom shutdown actions.

- If you use a Jetty-embedded web server instead of the default Tomcat, you will need to set the Servlet API to 5.0. Jetty does not yet support the Servlet API 6.0 used by default in Spring Boot 3.

- The Spring Framework 6 used by Spring Boot 3 removed the support for Apache HttpClient. If you used Apache HttpClient in your application in previous versions of Spring Boot, you may notice behavior changes.

As mentioned, you won't notice these changes unless your application explicitly relies on some of these features. However, there is a behavior change that may impact your application. In versions before Spring Boot 3, the URLs ending with a / would match with controllers without that trailing slash. For instance, GET /teams and GET /teams/ would match the same controller in our application. In Spring Boot 3 GET /teams/ will fail unless we prepare the application for that.

In this recipe, we'll prepare our application to manage the trailing slash of the requests as in previous versions to ensure that the clients relying on our application are not impacted by the Spring Boot upgrade.

Getting ready

We'll use the outcome of the *Managing Actuator changes* recipe for this recipe. I prepared a completed version in case you haven't done so yet. You may find it in the book's GitHub repository at https://github.com/PacktPublishing/Spring-Boot-3.0-Cookbook in the chapter9/recipe9-7/start folder.

How to do it...

Let's ensure our application consumers don't get disrupted by our upgrade to Spring Boot 3!

1. Let's add a custom web configuration. For that, create a file named WebConfiguration that implements the WebMvcConfigurer interface:

    ```
    @Configuration
    public class WebConfiguration implements WebMvcConfigurer
    ```

    Then, override the configurePathMatch method as follows:

    ```
    @Override
    public void configurePathMatch(PathMatchConfigurer configurer) {
        configurer.setUseTrailingSlashMatch(true);
    }
    ```

2. Now, you can validate the application's behavior by opening http://localhost:8080/teams/ in your browser. You can check that it works with or without a trailing slash.

3. If you run the application tests, you will realize that all of them have now succeeded, as only the tests related to trailing slash in the controller were still failing.

How it works...

Rather than relying on default behaviors that may change, the ending slash has been deprecated to enforce explicit matches for more stable and predictable applications as new versions appear.

Spring Boot provides the mechanism to maintain the same behavior as previous versions. However, you probably realized that the `setUseTrailingSlashMatch` method is deprecated. This is to warn developers about this non-recommended behavior and enforce the move to explicit matches.

There's more...

The same approach can be used with WebFlux. Instead of implementing `WebMvcConfigurer`, you would implement `WebFluxConfigurer`. It would look like this:

```
@Configuration
public class WebConfiguration implements WebFluxConfigurer {
    @Override
    public void configurePathMatching(PathMatchConfigurer configurer){
        configurer.setUseTrailingSlashMatch(true);
    }
}
```

See also

I recommend that you check the official guidelines for Spring Boot 3 migration at `https://github.com/spring-projects/spring-boot/wiki/Spring-Boot-3.0-Migration-Guide#web-application-changes`. As mentioned during the introduction of this recipe, if your application relies on some of the internal components impacted by the migration, you can find more details for fixing your issues.

Using OpenRewrite for migration automation

In this chapter, we made all migration upgrades manually. However, this approach can be too slow and error-prone in large code bases. Some tools try to automate this process, and the most popular one is OpenRewrite. OpenRewrite is a platform tool that aims to refactor any source code. According to its documentation, it aims to eliminate the technical debt of the developers' repositories. It provides a mechanism to run recipes for source code refactoring. A popular open source recipe tackles the subject of this chapter: Spring Boot migrations.

Getting ready

We'll use the original sample targeting Spring Boot 2.6.15 in this recipe. You can find it in the book's GitHub repository at `https://github.com/PacktPublishing/Spring-Boot-3.0-Cookbook` in the `chapter9/football` folder.

How to do it...

We'll do the same upgrade from Spring 2.6 to Spring 3.2.4, automating most of the process. Seven recipes in one!

1. Let's start by adding the OpenRewrite plugin to our project. To do that, add the following snippet to the plugins section of the pom.xml file:

```
<plugin>
    <groupId>org.openrewrite.maven</groupId>
    <artifactId>rewrite-maven-plugin</artifactId>
    <version>5.27.0</version>
    <configuration>
        <activeRecipes>
            <recipe>org.openrewrite.java.spring.boot2.
UpgradeSpringBoot_2_7</recipe>
        </activeRecipes>
    </configuration>
    <dependencies>
        <dependency>
            <groupId>org.openrewrite.recipe</groupId>
            <artifactId>rewrite-spring</artifactId>
            <version>5.7.0</version>
        </dependency>
    </dependencies>
</plugin>
```

When writing this book, the latest version of the OpenRewrite plugin is 5.7.0. Use the latest one when you try it.

We are using the OpenRewrite recipe to upgrade to Spring 2.7 because we'll do the upgrade gradually. In further steps, we'll upgrade to Spring Boot 3.0, then 3.1, and finally 3.2.

2. Next, we'll execute the OpenRewrite plugin. For that, open your terminal in the root folder of the project and execute the following Maven command:

```
./mvnw rewrite:run
```

You will see that it makes some changes:

- It upgrades the Spring Boot version to 2.7.18.

- It adds a test dependency to org.junit.jupiterter:junit-jupiter.

- It replaces the property spring.datasource.initialization-mode with spring.sql.init.mode in the application.yml file only. Take note of this change, as we'll need in the next step.

- It modifies the references to `org.junit` to `org.junit.jupiter` equivalents in the test classes.

3. Let's check if the upgrade to Spring Boot 2.7 works.

- If you run the tests, you will see that the tests that rely on Testcontainers don't work. This is because OpenRewrite excluded some required dependencies for Testcontainers. To fix it, open `pom.xml` and remove the `exclusion` `junit` on the Testcontainers dependency:

```
<dependency>
    <groupId>org.testcontainers</groupId>
    <artifactId>junit-jupiter</artifactId>
    <version>${testcontainers.version}</version>
    <scope>test</scope>
    <!-- <exclusions>
            <exclusion>
            <groupId>junit</groupId>
            <artifactId>junit</artifactId>
            </exclusion>
        </exclusions> -->
</dependency>
```

In this example, I just commented the `exclusions` section for clarity. In your code, you can remove it.

- Once the exclusion is fixed, you will see that it cannot load the application context because it cannot resolve the `spring.datasource.initialization-mode` configuration. To fix it, open the `CustomDatasourceService` class and modify the `@Value` annotation used in the constructor. You should replace the `spring.datasource.initialization-mode` setting with the `spring.sql.init.mode` setting.

- Rerun the tests and you will see that all of them succeed.

4. It's time to start with Spring Boot 3.0. To perform this upgrade, open the `pom.xml` file and modify `activeRecipes/recipe` to `org.openrewrite.java.spring.boot3.UpgradeSpringBoot_3_0`. The OpenRewrite plugin configuration should look like this:

```
<configuration>
    <activeRecipes>
        <recipe>org.openrewrite.java.spring.boot3.
UpgradeSpringBoot_3_0</recipe>
    </activeRecipes>
</configuration>
```

5.　Then, execute the plugin again. The changes performed by this upgrade are as follows:

- It replaced all `javax` references with `jakarta`.

- It migrated the Spring Security changes. It properly migrated `UserDetailsManager` and `WebSecurityCustomizer`. However, as we'll see in *step 7*, the `SecurityFilterChain` bean requires some adjustments.

- It migrated the settings defined in the `application.yml` file:

  - It replaced `management.trace.http.enabled` with `management.httpexchanges.recording.enabled`.

  - It migrated the `spring.data.cassandra` settings to `spring.cassandra`.

6.　In the Spring Boot 3.0 upgrade, the application doesn't compile. Let's fix the compilation errors:

- The `DataSourceInitializationMode` class has been removed in Spring Boot 3, but OpenRewrite has not migrated it. As we studied in the *Preparing the application* recipe, this change can be easily fixed by replacing `DataSourceInitializationMode` with `DatabaseInitializationMode`.

- Some Actuator changes were correctly applied, but others were not. The `application.yml` was correctly modified, replacing `management.trace.http.enabled` with `management.httpexchanges.recording.enabled`. However, `HttpTraceRepository` and `InMemoryHttpTraceRepository` were not migrated. You can replace them with `HttpExchangeRepository` and `InMemoryHttpExchangeRepository`. For more details, you can check out the *Managing Actuator changes* recipe.

- In the `details` field in the `MatchEventEntity` class, replace the annotation `@Type(JsonType.class)` with `@JdbcTypeCode(SqlTypes.JSON)`.

7.　Next, let's fix the tests:

- As explained in *step 3*, you should remove the `exclusion junit` from the Testcontainers dependency.

- As we no longer use the Hypersistence library, we can remove it from our project. For more details, check the *Upgrading Spring Data* recipe.

- OpenRewrite properly migrated the `spring.data.cassandra.*` settings to `spring.cassandra.*` in the `application.yml` file. However, it didn't modify the references in the tests. To fix it, just replace all references to `spring.data.cassandra` with `spring.cassandra`. See the *Upgrading Spring Data* recipe for more details.

- The MVC tests don't work. To fix them, we need to include the @Import(SecurityConfig.class) annotation in the FootballControllerTest and SecurityControllerTest classes. As mentioned previously, the SecurityFilterChain bean is migrated but requires some adjustments. As explained in the *Upgrade the project to Spring Boot 3* recipe, we need to switch the order of some calls. The method should look like this:

```
@Bean
SecurityFilterChain filterChain(HttpSecurity http)
                                            throws Exception {
    http.authorizeHttpRequests(requests -> requests
        .requestMatchers("/security/private/**").hasRole("ADMIN")
        .requestMatchers("/**").permitAll())
        .httpBasic(withDefaults());
    return http.build();
}
```

- OpenRewrite didn't migrate BigInteger to Long and the parameter order binding we studied in the *Upgrading Spring Data* recipe. To fix it, apply both changes explained in the *Upgrading Spring Data* recipe.

- The tests that check the trailing slash fail. If we want to keep the same behavior, we'll need to add a WebMvcConfigurer as explained in the *Managing web application changes* recipe.

After applying these fixes, both the tests and the application should work.

8. Next, let's upgrade to Spring Boot 3.1. For that, change the recipe in the pom.xml file to org.openrewrite.java.spring.boot3.UpgradeSpringBoot_3_1.

When writing this book, I found an issue related to the Testcontainers version when running this recipe. The message is similar to this:

```
[ERROR] Failed to execute goal org.openrewrite.maven:rewrite-
maven-plugin:5.27.0:run (default-cli) on project football:
Execution default-cli of goal org.openrewrite.maven:rewrite-
maven-plugin:5.27.0:run failed: Error while visiting chapter9/
recipe9-8/end/football/pom.xml: java.lang.IllegalStateException:
Illegal state while comparing versions : [1.19.7] and
[${testcontainers.version}.0.0.0.0.0.0]. Metadata = [null]
```

To avoid a runtime error, replace the testcontainers.version project variable in the pom.xml file with the real version in the Testcontainers dependencies. For instance, see this dependency:

```
<dependency>
    <groupId>org.testcontainers</groupId>
    <artifactId>junit-jupiter</artifactId>
    <version>${testcontainers.version}</version>
    <scope>test</scope>
</dependency>
```

Replace it with the following:

```
<dependency>
    <groupId>org.testcontainers</groupId>
    <artifactId>junit-jupiter</artifactId>
    <version>1.19.7</version>
    <scope>test</scope>
</dependency>
```

After executing the OpenRewrite plugin, you will see that only a few changes were applied to the pom.xml file. You will need to remove the exclusion junit on the Testcontainers dependency again.

9. In the Spring Boot 3.1 upgrade, there are no compilation errors. However, some tests fail. This can be fixed just by changing spring.jpa.database-platform to org.hibernate. dialect.PostgreSQLDialect as explained in the *Upgrading Spring Data* recipe. Rerun the test after applying this fix; all of them should succeed.

10. Lastly, upgrade to Spring Boot 3.2. For that, change the OpenRewrite recipe to org. openrewrite.java.spring.boot3.UpgradeSpringBoot_3_2. Again, you will need to remove the exclusion junit, but this time, no other actions are required. If you run the tests, they should succeed, and the application will run smoothly.

How it works...

OpenRewrite loads a representation of your code known as a **Lossless Semantic Trees** (**LSTs**) and then applies modifications to that representation by using **visitors**. Once the visitors are applied, OpenRewrite transforms the LSTs to code again. An OpenRewrite recipe is a set of visitors. For instance, one visitor changes the javax references to jakarta references in the LST, another visitor changes the Spring Data configuration settings, and so on, transforming the LSTs to the final upgraded version. Finally, the transformed LSTs are converted into code.

Usually, one OpenRewrite recipe definition does the migration of one Spring Boot version to the next one, e.g., from 2.6 to 2.7. The OpenRewrite Maven plugin detects all the recipes that it should apply from the current application version to the desired target version, and then it applies the recipes in order to make the upgrade gradual. For more information, see the *There's more* section.

As you might realize, in this recipe, many scenarios are not covered by the existing recipes. The OpenRewrite recipes are open source and maintained by the community. They handle the most common migration scenarios. For this chapter, I tried to prepare a sample with some scenarios that are not very rare but not very common, such as Hibernate scenarios using the BigInteger class. In any case, it's important to understand what changes are made to each upgrade so that if an error appears, we can fix it manually.

Having a good set of tests is always helpful, as they may help detect behavior changes between versions. In this chapter, we used extensive tests, specifically Testcontainers. They helped detect incompatibilities when accessing PostgreSQL and Cassandra.

There's more...

In this recipe, we made a gradual upgrade, but you can run the migration directly by applying the OpenRewrite `org.openrewrite.java.spring.boot3.UpgradeSpringBoot_3_2` recipe. You must apply the same additional fixes performed during the gradual migration:

- Before executing the `OpenRewrite` plugin, replace the Testcontainers version using a variable with a constant version.

- Remove the `exclusion junit` in the Testcontainers dependency.

- Replace `DataSourceInitializationMode` with `DatabaseInitializationMode`.

- Replace `HttpTraceRepository` and `InMemoryHttpTraceRepository` with `HttpExchangeRepository` and `InMemoryHttpExchangeRepository`.

- Replace the annotation `@Type(JsonType.class)` with `@JdbcTypeCode(SqlTypes.JSON)`.

- Replace all references to `spring.data.cassandra` with `spring.cassandra`.

- Add the `@Import(SecurityConfig.class)` annotation in the `FootballControllerTest` and `SecurityControllerTest` classes.

- Fix the `SecurityFilterChain` bean in the `SecurityConfig` class.

- Replace the `BigInteger` class with `Long` and the parameter order binding in the class in the `DynamicQueriesService` class.

- Add a `WebMvcConfigurer` if you want to keep the trailing slash behavior.

- Replace `org.hibernate.dialect.PostgreSQL82Dialect` with `org.hibernate.dialect.PostgreSQLDialect`.

See also

I recommend visiting the OpenRewrite website at `https://docs.openrewrite.org`. There are many recipes that can be used to maintain our code, not only for Spring Boot migrations.

Other tools aim to automate the migration process as much as possible. For instance, the Spring team developed an experimental project named Spring Boot Migrator. You can find more information at `https://github.com/spring-projects-experimental/spring-boot-migrator`.

Index

`packtpub.com`

Subscribe to our online digital library for full access to over 7,000 books and videos, as well as industry leading tools to help you plan your personal development and advance your career. For more information, please visit our website.

Why subscribe?

- Spend less time learning and more time coding with practical eBooks and Videos from over 4,000 industry professionals

- Improve your learning with Skill Plans built especially for you

- Get a free eBook or video every month

- Fully searchable for easy access to vital information

- Copy and paste, print, and bookmark content

Did you know that Packt offers eBook versions of every book published, with PDF and ePub files available? You can upgrade to the eBook version at `packtpub.com` and as a print book customer, you are entitled to a discount on the eBook copy. Get in touch with us at `customercare@packtpub.com` for more details.

At `www.packtpub.com`, you can also read a collection of free technical articles, sign up for a range of free newsletters, and receive exclusive discounts and offers on Packt books and eBooks.

Other Books You May Enjoy

If you enjoyed this book, you may be interested in these other books by Packt:

Modern API Development with Spring 6 and Spring Boot 3

Sourabh Sharma

ISBN: 978-1-80461-327-6

- Create enterprise-level APIs using Spring and Java.

- Understand and implement REST, gRPC, GraphQL, and asynchronous APIs for various purposes.

- Develop real-world web APIs and services, from design to deployment.

- Expand your knowledge of API specifications and implementation best practices.

- Design and implement secure APIs with authorization and authentication.

- Develop microservices-based solutions with workflow and orchestration engines.

- Acquire proficiency in designing and testing user interfaces for APIs.

- Implement logging and tracing mechanisms in your services and APIs

Learning Spring Boot 3.0

Greg L. Turnquist

ISBN: 978-1-80323-330-7

- Create powerful, production-grade web applications with minimal fuss.
- Support multiple environments with one artifact, and add production-grade support with features.
- Find out how to tweak your Java apps through different properties.
- Enhance the security model of your apps.
- Make use of enhancing features such as native deployment and reactive programming
- in Spring Boot.
- Build anything from lightweight unit tests to fully running embedded web container integration tests.
- Get a glimpse of reactive programming and decide if it's the right approach for you

Packt is searching for authors like you

If you're interested in becoming an author for Packt, please visit authors.packtpub.com and apply today. We have worked with thousands of developers and tech professionals, just like you, to help them share their insight with the global tech community. You can make a general application, apply for a specific hot topic that we are recruiting an author for, or submit your own idea.

Hi!

I am Felip Miguel Puig, author of *Spring Boot 3.0 Cookbook*. I really hope you enjoyed reading this book and found it useful for increasing your productivity and efficiency.

It would really help me (and other potential readers!) if you could leave a review on Amazon sharing your thoughts on this book.

Go to the link below or scan the QR code to leave your review:

`https://packt.link/r/1835089496`

Your review will help us to understand what's worked well in this book, and what could be improved upon for future editions, so it really is appreciated.

Best wishes,

Felip Miguel Puig

Download a free PDF copy of this book

Thanks for purchasing this book!

Do you like to read on the go but are unable to carry your print books everywhere?

Is your eBook purchase not compatible with the device of your choice?

Don't worry, now with every Packt book you get a DRM-free PDF version of that book at no cost.

Read anywhere, any place, on any device. Search, copy, and paste code from your favorite technical books directly into your application.

The perks don't stop there, you can get exclusive access to discounts, newsletters, and great free content in your inbox daily

Follow these simple steps to get the benefits:

1. Scan the QR code or visit the link below

https://packt.link/free-ebook/9781835089491

2. Submit your proof of purchase

3. That's it! We'll send your free PDF and other benefits to your email directly

www.ingramcontent.com/pod-product-compliance
Lightning Source LLC
Chambersburg PA
CBHW060649060326
40690CB00020B/4568